J. A. Bahturin, Lectures on Lie Algebras

STUDIEN ZUR ALGEBRA UND IHRE ANWENDUNGEN

Im Auftrag des
Zentralinstituts für Mathematik und Mechanik
der Akademie der Wissenschaften der DDR
herausgegeben von L. Budach und H.-J. Hoehnke

BAND 4

J. A. BAHTURIN

LECTURES ON LIE ALGEBRAS

J. A. BAHTURIN

LECTURES ON LIE ALGEBRAS

Lectures given at
Humboldt University Berlin and Lomonosov University Moscow

AKADEMIE-VERLAG · BERLIN
1978

Zentralinstitut für Mathematik und Mechanik, 108 Berlin, Mohrenstraße 39
Erschienen im Akademie-Verlag, 108 Berlin, Leipziger Str. 3—4

Lizenznummer: 202 · 100/406/78
Gesamtherstellung: VEB Druckerei „Thomas Müntzer", 582 Bad Langensalza
Bestellnummer: 762 492 9 (2174/4) · LSV 1025
Printed in GDR
DDR 28,– M

Preface

The present volume exhibits that part of the Lie algebra theory which develops under the fruitful influence of the infinite group theory or is destined for certain purposes of this latter.

As the reader may easily observe from the table of contents, our account concentrates around the concept of identical relation either in the Lie algebras themselves or in the related algebraic objects. Varieties of Lie algebras are dealt with in chapters 3 and 5, miscellaneous properties of Lie algebras with identities connected primarily with the concept of the universal envelope is the concern of chapter 4. The above mentioned topics need many preliminaries (chapter 1) and a good theory of free Lie algebras (chapter 2). The first five chapters being a demonstration of the heuristic role of group theory in the theory considered, chapter 6 effects a feedback in the interrelations of these two theories. Namely, in this chapter we apply the results hitherto obtained to clarify the commutator structure of certain relatively free groups.

The exposition is selfcontained, the only knowledge required is some elementary facts from the group, ring and module theories. The reference, e. g., 'lemma 2,3.1' means the lemma 1 from the section 3 of the chapter 2.

The book is based on the lectures given at Humboldt and Lomonosov Universities (1973—1975); it is a pleasure for me to thank the staffs of the chairs of algebra of the two Universities for their friendly assistance. I am indebted especially to Professors A. I. Kostrikin and A. L. Šmel'kin for their advices and useful discussion, to Dr. A. Ju. Ol'šanskiĭ for the cooperation and to Professors A. L. Skornjakov and H.-J. Hoehnke who suggested me to publish my lectures.

Table of contents

Chapter 1. Basic concepts and results 1

1.1. The definition of a Lie algebra 1
1.2. Examples 2
1.3. Isomorphisms, homomorphisms, subalgebras, ideals, factor algebras 3
1.4. Derivations 4
1.5. Direct, cartesian and semidirect products of Lie algebras 4
1.6. Some remarkable series in a Lie algebra 6
1.7. Nilpotent algebras 7
1.8. Generators of nilpotent algebras 9
1.9. Engel's theorem 9
1.10. Two criteria for the nilpotency 11
1.11. Solvable Lie algebras 12
1.12. Higgins' theorem 12
1.13. The Frattini subalgebra 13
1.14. Restricted Lie algebras 16
1.15. Modules 18

Chapter 2. Free Lie algebras 22

2.1. The definition of a free Lie algebra 22
2.2. Bases of free Lie algebras 24
2.3. Subalgebras of free Lie algebras 28
2.4. The universal enveloping algebra 31
2.5. Free restricted Lie algebras 33
2.6. Free products 36

Chapter 3. Varieties of Lie algebras 39

3.1. Identical relations in Lie algebras 39
3.2. Verbal subalgebras 41
3.3. Linearization of identities 42
3.4. Lie algebra varieties over infinite fields 43
3.5. Product varieties 45
3.6. Some infinitely based varieties of Lie algebras 49
3.7. Wreath products 53
3.8. Applications to solvable varieties 55

Chapter 4. Universal enveloping algebra. Further results and applications 59

4.1. PI-algebras 59
4.2. Modules and identities. Zero characteristic case 61
4.3. Modules and identities. Prime characteristic case 63

4.4. Some centralizer concerned techniques . . . 65
4.5. Centralizers of elements in Lie algebras whose universal envelopes are PI-algebras . . . 67
4.6. Inner derivations in Lie algebras whose universal envelopes are PI-algebras . . . 71
4.7. Main theorem on polynomial identities in universal envelopes of Lie algebras . . . 72
4.8. Residual properties of Lie algebras . . . 73
4.9. Finitely generated algebraic Lie algebras . . . 76

Chapter 5. Identities in finite Lie algebras . . . 80

5.1. A characterization of a variety generated by a finite Lie algebra 80
5.2. Critical algebras . . . 81
5.3. Minimal representation . . . 83
5.4. Critical algebras in a class $\mathfrak{C}(f, m, c)$ . . . 84
5.5. Varieties $\mathfrak{B}^{(n)}$ . . . 86
5.6. Chief centralizers and maximal subvarieties . . . 89

Chapter 6. Applications to the group theory . . . 92

6.1. Topological algebras . . . 92
6.2. Campbell-Hausdorff formula . . . 95
6.3. Free groups . . . 97
6.4. Commutator calculus . . . 99
6.5. Nilpotent groups . . . 102
6.6. Malcev completion . . . 104
6.7. Isomorphism of categories of nilpotent Lie Q-algebras and nilpotent D-groups . . . 107
6.8. Relatively free groups . . . 109
6.9. Product varieties of groups . . . 112
6.10. Magnus groups . . . 115

Sources. References. Discussion . . . 117

Author Index . . . 123

Subject Index . . . 124

CHAPTER 1

Basic concepts and results

1.1. The definition of a Lie algebra

Let R be an associative ring in which the multiplication is written through juxtaposition. Consider another binary operation in R denoted by bracketing and derived from the former operation by the rule

$$[x, y] = xy - yx , \qquad x, y \in R .$$

The new operation in R possesses *inter alia* two properties:

(i) $[x, x] = 0$ for any $x \in R$;

(ii) $[x, [y, z]] + [y, [z, x]] + [z, [x, y]] = 0$ for any $x, y, z \in R$.

The first of these relations being trivial, we verify the second one:

$$[x, [y, z]] = [x, yz - zy] = xyz - xzy - yzx + zyx .$$

Similarly,

$$[y, [z, x]] = yzx - yxz - zxy + xzy , \quad [z, [x, y]] = zxy - zyx - xyz + yxz .$$

Adding these three sums one obtains zero, which proves (ii).

The identical relation $[x, x] = 0$ implies the so-called "anticommutativity identity":

$$[x, y] = - [y, x] .$$

The identical relation (ii) is called the *Jacobi identity*. The left hand side of (ii) is sometimes called the "jacobian" of x, y, z and denoted by $J(x, y, z)$.

Note without proof that all the "commutator identities", i.e. the identities holding for the *commutator* $[x, y]$, $x, y \in R$, R being an arbitrary associative ring, are consequences of $[x, x] \equiv 0$ and $J(x, y, z) \equiv 0$.

This (and many other reasons) justifies the introduction of a new class of rings. Namely, by a *Lie ring* we shall mean a ring A in which the multiplication (often denoted by [,]) is subject to the above mentioned identities:

$$[x, x] \equiv 0 \quad \text{and} \quad J(x, y, z) \equiv 0 . \tag{1}$$

If the underlaying abelian group A is endowed with a k-module structure, k being some commutative ring with unity 1, and

$$[\lambda x, y] = [x, \lambda y] = \lambda[x, y] , \qquad \lambda \in k, \; x, y \in A , \tag{2}$$

then A is called a *Lie algebra* over k. Thus any Lie ring is a Lie algebra over the ring $\mathbf{Z}$ of rational integers.

Note that there are "many" modules over some rings which can be turned into Lie algebras over the same rings only by putting $[x, y] = 0$ for all x, y. For example, let k be a ring without zero divisors, A be its ideal, x, y be nonzero elements from A. Then

$$0 = [yx, yx] = y[x, yx] = y[x, xy] = yx[x, y] .$$

Clearly then $[x, y] = 0$ for all $x, y \in A$.

On the other hand putting $[x, y] = 0$ in any k-module A for any k turns A into a Lie algebra over k. A Lie algebra with this property is called an *abelian Lie algebra* over k.

1.2. Examples

(i) Let $\mathbf{R}^3$ be the three-dimensional linear space over the field of real numbers in which $[\bar{u}, \bar{v}] = \bar{u} \times \bar{v}$, the vector product of $\bar{u}$ and $\bar{v}$, for any two vectors $\bar{u}$ and $\bar{v}$. It is well known that then 1.(1) and 1.(2) hold, so that we obtain a (nonabelian) Lie algebra over $\mathbf{R}$.

(ii) Let A be an associative and commutative k-algebra, k being some ring, and D being some *derivation* of A, that is a k-linear map $D: A \to A$ such that

$$D(ab) = D(a)\, b + aD(b) , \qquad a, b \in A .$$

we define an operation in A by the formula

$$[a, b] = aD(b) - D(a)\, b , \qquad a, b \in A .$$

Then (the proof is left to the reader) A becomes a Lie algebra over k. For example, let k be a field of characteristic $p > 0$, $A = k[x]$ be a polynomial ring over k, $D(f(x)) = f'(x)$, $B = A/(x^p)$, $\bar{D}: \bar{f} \to \overline{D(f)}$ be the derivation of B induced by D. Then under the commutator

$$[\bar{f}, \bar{g}] = \bar{f}\bar{g}' - \bar{f}'\bar{g}$$

B becomes a Lie algebra over k called a *Witt algebra* and denoted by W_1.

(iii) Let k_n be the algebra of all n by n matrices over k, $sl_n(k)$ be its submodule consisting of all matrices of trace zero. Then $sl_n(k)$ is closed under the commutator of two matrices and thus is a Lie algebra over k.

(iv) Let G be a group. Denote by $\gamma_s(G)$, $s = 1, 2, \ldots$, the members of the lower central series of G (see all the definitions and necessary results in the Chapter 6). Then $\gamma_s(G)$ is a normal subgroup of G and $L_s(G) = \gamma_s(G)/\gamma_{s+1}(G)$, $s = 1, 2, \ldots$, is an abelian group. Form a direct sum $L(G) = \bigoplus_{s=1}^{\infty} L_s(G)$. To turn $L(G)$ into a Lie ring it suffices to define $[\xi, \eta]$, $\xi \in L_s(G)$, $\eta \in L_r(G)$, since any element from $L(G)$ is a finite sum of its components from $L_i(G)$, $i = 1, 2, \ldots$. But $\xi = x\gamma_{s+1}(G)$, $\eta = y\gamma_{r+1}(G)$. Then put $[\xi, \eta] = ((x, y))\, \gamma_{s+r+1}(G)$, $((x, y))$ being a commutator of elements x, y in G. Using Hall's commutator identities (6, 4.) one readily verifies that the definition is unambiguous and satisfies the identities 1.(1)

For example, let G be the quaternion group (consisting of ± 1, $\pm i$, $\pm j$, $\pm k$ with common multiplication $i^2 = j^2 = k^2 = -1$, $ij = -ji = k$, $jk = -kj = i$, $ki = -ik = j$ subject to the rule of "signs"). Then $\gamma_1(G) = G$, $\gamma_2(G) = \{\pm 1\}$, $\gamma_3(G) = \{1\}$ and $L_1(G) = \{\bar{i}\}_2 \oplus \{\bar{j}\}_2$, $L_2(G) = \{\overline{-1}\}_2$, $L_3(G) = \{0\}$. Thus

$$L(G) = \{\bar{i}\}_2 \oplus \{\bar{j}\}_2 \oplus \{\overline{-1}\}_2 ,$$

all nonzero elements having order 2 and $[\bar{i}, \bar{j}] = \overline{-1}$, $[\bar{i}, \overline{-1}] = [\bar{j}, \overline{-1}] = 0$.

1.3. Isomorphisms, homomorphisms, subalgebras, ideals, factor algebras

To identify and to compare different Lie algebras by their algebraic properties one uses the concepts of *isomorphism* and *homomorphism* of Lie algebras over the same ring k. These are respectively the k-module isomorphism and homomorphism φ such that for any two elements x and y one has $\varphi([x, y]) = [\varphi(x), \varphi(y)]$. As usually, if φ is an isomorphism of A onto B then A is said to be *isomorphic* with B, denoted by $A \cong B$.

Given a k-submodule B of a Lie algebra A over k, call it a *subalgebra* (respectively an *ideal*) of A if for any $x \in B$ and any $y \in B$ (respectively $y \in A$) one has $[x, y] \in B$.

Let X be a subset of a Lie algebra A, B be a subalgebra of A. X is called a *generating set* of B if the least subalgebra of A containing X coincides with B. This fact is denoted by $B = \text{alg}(X)$. If B is the least ideal of A containing X then we say that B is *generated* by X *as an ideal of* A, this fact being denoted by $B = \text{id}_A(X)$.

Given an ideal B of A (this is commonly denoted by $B \trianglelefteq A$), the factor module A/B can be endowed with a multiplication through

$$[x + B, y + B] = [x, y] + B\,, \qquad x, y \in A\,.$$

One readily verifies that this formula gives a multiplication of cosets and thus obtains a Lie algebra which is called a *quotient algebra of* A *by* B, denoted by A/B.

Finally, given a subalgebra B of some algebra A and an ideal C of B, the quotient algebra B/C is called a *factor* of the algebra A. The factor B/C is called a *chief* one if C, B are ideals of A and B/C is a minimal nonzero ideal of A/C.

The following theorem collects some necessary facts connecting the just given definitions. However we omit proofs for they can be obtained by simple translations of the corresponding module- (or ring-) theoretical assertions.

Theorem 1. (i) *Let φ be a homomorphism of a Lie algebra A into a Lie algebra B. Then its kernel* Ker φ *is an ideal of A, its image* Im φ *is a subalgebra of B and $A/\text{Ker}\,\varphi \cong \text{Im}\,\varphi$.*

(ii) *Let A be a Lie algebra, B be its subalgebra, C be its ideal. Then $B + C$ is a subalgebra of A and $B + B/C \cong B/B \cap C$.*

(iii) *Let A be a Lie algebra, B, C be its ideals, $B \supseteq C$, then B/C is an ideal of A/C and $(A/C)/(B/C) \cong A/B$.*

A subalgebra (ideal) of a Lie algebra A is called a *proper* subalgebra (ideal) if $B \neq A$. A nonabelian algebra A is called *simple* if A contains no proper nonzero ideals. For example, the algebras $\mathbf{R}^3$, W_1 (for $p \geqq 3$) and $sl_n(k)$, if $n \geqq 2$ and k is a field of characteristic zero, are simple. We prove this for the case of W_1, the rest cases being left to the reader as exercises.

So far, denote by b_i the images of x^i, $i = 0, 1, 2, \ldots$, in $B = W_1$. Then $b_j = 0$, $j \geqq p$, $b_0, \ldots, b_{p-1}$ are linearly independent and the multiplication in B is given by the rule $[b_i, b_j] = (j - i)\, b_{i+j-1}$, $0 \leqq i, j \leqq p - 1$. Let K be a nonzero ideal of B. If $b_1 \in K$, then $K \ni [b_1, b_j] = (j - 1)\, b_j$, so that $b_0, b_2, \ldots, b_{p-1} \in K$ and therefore $K = B$. Given further a nonzero element $b = \alpha_1 b_{n_1} + \alpha_2 b_{n_2} + \ldots + \alpha_s b_{n_s} \in K$ with minimal s, assume $s \geqq 2$. Then multiplying b suitably many times by b_2 one has that $b_{p-1} \in K$ which contradicts our assumption. Hence K contains either b_0 and, consequently $b_1 = \frac{1}{2}[b_0, b_2]$, or b_s, $p > s \geqq 2$, and consequently $b_1 = \dfrac{1}{s!}[b_0, [b_0, \ldots, [b_0, b_s] \ldots]]$. In both cases $K = B$.

1.4. Derivations

Recall (2.(ii)) that a k-linear map d of some (not necessarily Lie or associative) algebra A into itself is called a *derivation* of A if $d([x, y]) = [d(x), y] + [x, d(y)]$, x, y being some arbitrary elements from A. If A is an abelian Lie algebra over k then any k-linear map of A into itself is a derivation of A. Denote by $\mathrm{End}_k(M)$ the set of all k-linear maps of some k-module M into itself. Then $\mathrm{End}_k(M)$ is an associative and thus a Lie algebra over k. Let A be a Lie algebra. The set of all derivations of A is a subset of $\mathrm{End}_k(A)$ and is denoted by $\mathrm{Der}_k(A)$. A derivation $d \in \mathrm{Der}_k(A)$ is called an *inner* derivation if there exists a $y \in A$ such that $d(x) = [y, x]$ for any $x \in A$. Conversely, given an element $z \in A$, consider a map $\mathrm{ad}(z)\colon A \to A$ defined by $(\mathrm{ad}(z))(x) = [z, x]$ for any $x \in A$. Clearly $\mathrm{ad}(z) \in \mathrm{End}_k(A)$. We show that $\mathrm{ad}(z) \in \mathrm{Der}_k(A)$:

$$\begin{aligned}(\mathrm{ad}(z))([x, y]) &= [[z, [x, y]] = [z, [x, y]] - J(x, y, z) = \\ &= -[x, [y, z]] - [y, [z, x]] = [x, [z, y]] + [[z, x], y] = \\ &= [(\mathrm{ad}(z))(x), y] + [x, (\mathrm{ad}(z))(y)] \,.\end{aligned}$$

The set of all inner derivations of A is denoted by $\mathrm{ad}(A)$. We distinguish the set of all $z \in A$ such that $\mathrm{ad}(z) = 0$, or, equivalently, $[z, x] = 0$ for all $x \in A$, as the *centre* of A and denote it by $Z(A)$.

Theorem 2. *Let A be a Lie algebra over some ring k.*

(i) $\mathrm{Der}_k(A)$ *is a Lie subalgebra of the Lie algebra* $\mathrm{End}_k(A)$;

(ii) $\mathrm{ad}(A)$ *is an ideal of* $\mathrm{Der}_k(A)$;

(iii) *the mapping* $x \mapsto \mathrm{ad}(x)$ *is a surjective homomorphism of A onto* $\mathrm{ad}(A)$ *whose kernel equals to the centre $Z(A)$ of A.*

Proof. (i) Let $d_1, d_2 \in \mathrm{Der}_k(A)$. We show that $[d_1, d_2] = d_1d_2 - d_2d_1 \in \mathrm{Der}_k(A)$. Indeed, $([d_1, d_2])([x, y]) = (d_1d_2 - d_2d_1)([x, y]) = d_1([d_2(x), y] + [x, d_2(y)]) - d_2([d_1(x), y] + [x, d_1(y)]) = [(d_1d_2)(x), y] + [d_2(x), d_1(y)] + [d_1(x), d_2(y)] + [x, (d_1d_2)(y)] - [(d_2d_1)(x), y] - [d_2(x), d_1(y)] - [d_1(x), d_2(y)] - [x, (d_2d_1)(y)] = [(d_1d_2 - d_2d_1)(x), y] + [x, (d_1d_2 - d_2d_1)(y)] = [([d_1, d_2])(x), y] + [x, ([d_1, d_2])(y)]$.

(ii) Let $d \in \mathrm{Der}_k(A)$, $\mathrm{ad}(x) \in \mathrm{ad}(A)$, $y \in A$. Then $[d, \mathrm{ad}(x)](y) = (d \cdot \mathrm{ad}(x) - \mathrm{ad}(x) \cdot d)(y) = d([x, y]) - [x, d(y)] = [d(x), y] + [x, d(y)] - [x, d(y)] = (\mathrm{ad}(d(x)))(y)$, thus $[d, \mathrm{ad}(x)] = \mathrm{ad}(d(x)) \in \mathrm{ad}(A)$ and (ii) is completed.

(iii) Let $x, y, z \in A$, then $(\mathrm{ad}([x, y]))(z) = [[x, y], z] = -[z, [x, y]] + J(x, y, z) = [x, [y, z]] + [y, [z, x]] = (\mathrm{ad}(x))(\mathrm{ad}(y))(z) - (\mathrm{ad}(y))(\mathrm{ad}(x))(z) = [\mathrm{ad}(x), \mathrm{ad}(y)](z)$, which proves that $x \mapsto \mathrm{ad}(x)$ is a homomorphism. Clearly $Z(A) \subseteq \mathrm{Ker}(\mathrm{ad})$. Let also $x \in \mathrm{Ker}(\mathrm{ad})$. Then $\mathrm{ad}(x) = 0$ and $(\mathrm{ad}(x))(y) = [x, y] = 0$ for any $y \in A$, so that $x \in Z(A)$.

Thus the proof of the theorem is complete.

Corollary. *$Z(A)$ is an ideal of A such that $A/Z(A) \cong \mathrm{ad}(A)$.*

1.5. Direct, cartesian and semidirect products of Lie algebras

Given a family $(A_\alpha)_{\alpha \in I}$, I being some index set, of Lie algebras over some ring k, consider their k-module cartesian product $A = C\prod_{\alpha \in I} A_\alpha$. Recall that A is the set of all functions $f: I \to \bigcup_{\alpha \in I} A_\alpha$ such that $f(\alpha) \in A_\alpha$ with componentwise addition $(f + g)(\alpha) = f(\alpha) + g(\alpha)$ and multiplication by scalars from k, $(\lambda f)(\alpha) = \lambda f(\alpha)$, $f, g \in A$, $\lambda \in k$, $\alpha \in I$. By putting $[f, g](\alpha) = [f(\alpha), g(\alpha)]$ we turn A into a Lie algebra called a *cartesian product* of a family $(A_\alpha)_{\alpha \in I}$. Denoting by $\pi_\alpha: A \to A_\alpha$ the projection of A onto A_α, the subalgebra $D \subseteq A$ is called a subcartesian product of the family A_α, $\alpha \in I$, if $\pi_\alpha|_D$ is onto for each $\alpha \in I$. If $f \in A$ then supp(f) is the subset of I consisting of all α such that $f(\alpha) \neq 0$. One readily verifies that the set B of functions f from A with finite support supp(f) is an ideal of A called a *direct product* of A_α, $\alpha \in I$, denoted by $\prod_{\alpha \in I} A_\alpha$.

The cartesian product A contains ideals $B_\alpha \cong A_\alpha$ for each $\alpha \in I$. These ideals are formed by all functions f such that $f(\beta) = 0$ for any $\beta \neq \alpha$. One easily sees that B is a direct sum of its submodules B_α, that is $B = \sum_{\alpha \in I} B_\alpha$ and $B_\alpha \cap \sum_{\beta \neq \alpha} B_\beta = \{0\}$. Conversely, if some algebra B is a direct sum of its submodules A_α, $\alpha \in I$, which are its ideals then B is isomorphic with the direct product $\prod_{\alpha \in I} A_\alpha$, the isomorphisms being induced by identity mappings of each A_α, $\alpha \in I$. Therefore, if some algebra is a direct module sum of a set of its ideals we call it a direct product of these latter.

When $I = \{1, 2, \ldots, n\}$, we denote $\prod_{i=1}^{n} A_i$ $(= C\prod_{i=1}^{n} A_i)$ also by $A_1 \times A_2 \times \ldots \times A_n$. A direct product of two algebras is a particular case of some more general construction called a *semidirect product* of Lie algebras.

Given two Lie algebras A and B over k and a homomorphism $\pi: A \to \mathrm{Der}_k(B)$, consider their module direct product $A \times B$ which we now identify with the set of all pairs (a, b), $a \in A$, $b \in B$, with componentwise addition and multiplication by scalars. Put

$$[(a_1, b_1), (a_2, b_2)] = \big([a_1, a_2], (\pi(a_1))(b_2) - (\pi(a_2))(b_1) + [b_1, b_2]\big) . \tag{1}$$

The algebra obtained from A and B subject to this operation is called a semidirect product of A and B with respect to π and denoted by $A \curlywedge B$.

Theorem 3. (i) *$A \curlywedge B$ is a Lie algebra possessing a subalgebra $\overline{A} \cong A$ and an ideal $\overline{B} \cong B$ such that $A \curlywedge \overline{B} = \overline{A} \oplus B$.*

(ii) *Conversely, let $C = A \oplus B$, A being a subalgebra of C, B being its ideal (a split extension of A by B). Let also $\pi(a) = (\mathrm{ad}(a))|_B$ be the restriction of* ad(a), $a \in A$, *on B. Then C is isomorphic with a semidirect product $A \curlywedge B$ with respect to π.*

Proof. (i) The first of the set of laws 1.(1) being trivially fulfilled for $A \curlywedge B$, it suffices to verify that $J(u, v, w) = 0$ for all $u, v, w \in A \curlywedge B = C$. Moreover, since $J: C \times C \times C \to C$ is a skew-symmetrical trilinear map and since $(a, b) = (a, 0) + (0, b)$, the arguments u, v, w may be assumed to have the form $(a, 0)$ or $(0, b)$ for some $a \in A$, $b \in B$, and ordered. Then only four cases may occur:

$$J\big((a_1, 0), (a_2, 0), (a_3, 0)\big) = 0\,, \qquad J\big((a_1, 0), (a_2, 0), (0, b_3)\big) = 0\,,$$

$$J\big((a_1, 0), (0, b_2), (0, b_3)\big) = 0\,, \qquad J\big((0, b_1), (0, b_2), (0, b_3)\big) = 0\,.$$

But the first and the last relations may be rewritten in the form $(J(a_1, a_2, a_3), 0) = 0$ and $(0, J(b_1, b_2, b_3)) = 0$ which hold since A and B are Lie algebras. The rest two relations will be dealt with separately.

$$J((a_1, 0), (a_2, 0), (0, b_3)) = (0, (\pi(a_1))((\pi(a_2))(b_3)) +$$
$$+ (0, (\pi(a_2))((-\pi(a_1))(b_3))) + (0, (-\pi([a_1, a_2]))(b_3)) =$$
$$= (0, (\pi(a_1)\pi(a_2) - \pi(a_2)\pi(a_1) - \pi([a_1, a_2]))(b_3)) = 0$$

since π is a homomorphism of A into $\mathrm{Der}_k(B)$.

$$J((a_1, 0), (0, b_2), (0, b_3)) =$$
$$= (0, (\pi(a_1))([b_2, b_3])) + (0, [b_2, (-\pi(a_1))(b_3)]) + (0, [b_3, (\pi(a_1))(b_2)]) =$$
$$= (0, (\pi(a_1))([b_2, b_3]) - [b_2, (\pi(a_1))(b_3)] - [(\pi(a_1))(b_2), b_3]) = 0\,,$$

since $\pi(a_1)$ is a derivation of B. The second part of (i) is clear if one considers a set $\bar{A}$ of pairs $(a, 0)$, $a \in A$, and $\bar{B}$, whose elements are $(0, b)$, $b \in B$.

(ii) Given an element $x = a + b \in C$, put $\varphi(x) = (a, b) \in A \curlywedge B$. Obviously, φ is a k-linear isomorphism of modules C and $A \curlywedge B$. Let $x_1 = a_1 + b_1$, $x_2 = a_2 + b_2$. Then $[x_1, x_2] = [a_1 + b_1, a_2 + b_2] = [a_1, a_2] + [a_1, b_2] - [a_2, b_1] + [b_1, b_2] = [a_1, a_2] + (\pi(a_1))(b_2) - (\pi(a_2))(b_1) + [b_1, b_2]$. From the definition of φ and (1) we immediately observe that $\varphi([x_1, x_2]) = [\varphi(x_1), \varphi(x_2)]$, and we are finished.

Examples. (i) Let M be a k-module, A be Lie a subalgebra of $\mathrm{End}_k M$. Then M can be considered as an abelian Lie algebra, so that $\mathrm{Der}_k M = \mathrm{End}_k M$. This enables us to form a semidirect product $P = A \curlywedge M$ with respect to the identity map $A \to \mathrm{Der}_k M$. For example, let $M = k[t]$ be a polynomial ring over k, A be a submodule of $\mathrm{End}_k M$ spanned by transformations e_1, e_2, e_3 such that $e_1(f(t)) = \frac{\mathrm{d}f}{\mathrm{d}t}$, $e_2(f(t)) = tf(t)$, $e_3(f(t)) = f(t)$. One readily verifies that $[e_1, e_2] = e_3$, $[e_1, e_3] = [e_2, e_3] = 0$ so that A is a Lie subalgebra of $\mathrm{End}_k M$ and one can form a semidirect product $P = A \curlywedge M$. If k is a field of characteristic zero, any ideal of P either is $\{0\}$ or contains M.

(ii) Let S be a nonabelian two-dimensional Lie algebra over some field k. Given two linearly independent elements $x, y \in S$, consider $[x, y] = \alpha x + \beta y$, $\alpha, \beta \in k$. Since $[x, y] \neq 0$, one of the coefficients, say α, is nonzero. Put then $e = [x, y]$, $f = -\alpha^{-1}y$. Then we have a basis of S with the multiplication table $[f, e] = e$. Putting, further, $A = kf$, $B = ke$, one obtains a subalgebra A and an ideal B such that $S = A \oplus B$. By the second part of the theorem 2, $S \cong A \curlywedge B$.

1.6. Some remarkable series in a Lie algebra

Let A be a Lie algebra over some ring k, B, C be its submodules. By $[B, C]$ we denote a submodule of A spanned by the commutators $[b, c]$, $b \in B$, $c \in C$. Clearly, $[C, B] = [B, C]$. If B and C are ideals of A then so does $[B, C]$. If $\varphi\colon A \to D$ is a homomorphism of Lie algebras then $\varphi([B, C]) = [\varphi(B), \varphi(C)]$.

We define now three especially useful series of ideals of a Lie algebra A.

(i) *Derived series.* Put $A' = [A, A]$, and if $A', A'', \dots, A^{(s)}$ are already defined, put $A^{(s+1)} = [A^{(s)}, A^{(s)}]$. The ideals $A^{(s)}$, $s = 1, 2, \dots$, form a (possibly non strictly) decreasing series

$$A = A^{(0)} \supseteq A^{(1)} = A' \supseteq A^{(2)} = A'' \supseteq \dots \supseteq A^{(s)} \supseteq A^{(s+1)} \supseteq \dots$$

whose subsequent factors $A^{(s)}/A^{(s+1)}$ are abelian algebras, $s = 0, 1, 2, \dots$.

(ii) *Lower central series.* Put $A^1 = A$. Given A^s, for some $s \geqq 1$, put $A^{s+1} = [A, A^s]$. Then

$$A = A^1 \supseteq A^2 \supseteq \dots \supseteq A^s \supseteq A^{s+1} \supseteq \dots$$

is a (possibly non strictly) descending series of A whose subsequent factors A^s/A^{s+1} are central in A/A^{s+1}. The common member $A' = A^2$ of both derived and lower series of A is called a *derived algebra* of A.

Note, that the remark from the first paragraph of this section yields that if φ is an epimorphism of A onto B then $\varphi(A^{(s)}) = (\varphi(A))^{(s)} = B^{(s)}$ and $\varphi(A^s) = (\varphi(A))^s = B^s$ for all $s = 0, 1, 2, \dots$.

(iii) *Upper central series.* Put $Z_0(A) = \{0\}$ and define $Z_s(A)$, $s \geqq 1$, inductively and so that $Z_{s+1}(A) \supseteq Z_s(A)$ by the rule: $Z_{s+1}(A)/Z_s(A)$ is the centre of $A/Z_s(A)$. It is clear that $Z_s(A)$ is an ideal of A for any $s \geqq 0$ and that $Z_1(A) = Z(A)$, the centre of A.

Remark, that if $d \in \mathrm{Der}_k(A)$, then $d(A^{(s)}) \subseteq A^{(s)}$, $d(A^s) \subseteq A^s$ and $d(Z_s(A)) \subseteq Z_s(A)$ which can be proved by direct calculations. The subalgebras invariant under all derivations of the embracing algebras are termed *characteristic.*

Examples. (i) Let A be the two-dimensional nonabelian Lie algebra $A = ke \curlywedge kf$. Then $A' = kf$ and $A'' = \{0\}$. Quite on the contrary $A^s = A^2 \neq \{0\}$ for all $s \geqq 2$. The upper central series of A is trivial since $Z(A) = \{0\}$.

(ii) Let A be the Lie algebra of all matrices of the form

$$\begin{pmatrix} 0 & \alpha & \beta \\ 0 & 0 & \gamma \\ 0 & 0 & 0 \end{pmatrix}, \quad \alpha, \beta, \gamma \in k\,.$$

Then the derived algebra A' of A consists of matrices

$$\begin{pmatrix} 0 & 0 & \alpha \\ 0 & 0 & 0 \\ 0 & 0 & 0 \end{pmatrix}, \quad \alpha \in k\,.$$

Therefore, $A'' = A^3 = \{0\}$. The members of the upper central series of A are easily seen to be $Z_1(A) = A'$, $Z_2(A) = A$.

1.7. Nilpotent algebras

Proposition 1. *Let A be a Lie algebra over some ring k such that for some positive integer c one has $A^c \neq \{0\}$ but $A^{c+1} = \{0\}$. Then $Z_{c-1}(A) \neq A$ but $Z_c(A) = A$. The converse is also true.*

Proof. Show that $A^l \subseteq Z_{c-l+1}(A)$. This is vacuously true for $l = c + 1$. Assume then that $A^{l+1} \subseteq Z_{c-l}(A)$. Then for each $x \in A$, $y \in A^l$ one has $[x, y] \in A^{l+1} \subseteq Z_{c-l}$. But then $[x + Z_{c-l}, y + Z_{c-l}] = 0$ for any $x + Z_{c-l} \in A/Z_{c-l}$. By the definition then

$x \in Z_{c-l+1}$ and therefore $A^l \subseteq Z_{c-l+1}$. In particular, $Z_c = Z_c(A) = A$. Moreover, if $Z_{c-1}(A) = A$, then quite in the same way $Z_{c-l} \supseteq A^l$, $l = 1, 2, \ldots$, and thus $A^c = \{0\}$ which is inconsistently.

A Lie algebra A satisfying the equivalent conditions of the just proved proposition is called a *nilpotent* Lie algebra of the *nilpotency class c*. For example, a nonabelian two-dimensional Lie algebra over some field k is not nilpotent. On the contrary, the Lie algebra of example (ii) is nilpotent of class 2. Moreover, given a Lie algebra of upper nil triangular matrices of order m, $nt_m(k)$, one easily observes that it is nilpotent of class $m - 1$.

The property of a Lie algebra to be nilpotent has a number of alternative equivalent forms. For example, given some positive integer n, some arrangement σ of brackets in the word $x_1 x_2 \ldots x_n$, some set $\{a_1, a_2, \ldots, a_n\}$ of elements of A, denote by $[a_1, a_2, \ldots, a_n]_\sigma$ the product of $a_1, a_2, \ldots, a_n$ subject to the arrangement of brackets σ. If σ is a "left normed" arrangement, that is, $[x_1, [x_2, \ldots, [x_{n-1}, x_n] \ldots]]$, then $[a_1, a_2, \ldots, a_n]_\sigma$ is denoted simply by $[a_1, a_2, \ldots, a_n]$.

Proposition 2. *The following three properties of a Lie algebra A over some ring k are equivalent:*

(i) *A is nilpotent of class c;*

(ii) *c is the least positive integer such that for any arrangement σ of brackets and any elements $a_1, a_2, \ldots, a_{c+1}$ (not necessarily distinct) from A,* $[a_1, a_2, \ldots, a_{c+1}]_\sigma = 0$;

(iii) *c is the least positive integer such that for any $a_1, a_2, \ldots, a_{c+1} \in A$,* $[a_1, a_2, \ldots, a_{c+1}] = 0$.

Proof. From the Jacobi identity one easily deduces that $[A^s, A^r] \subseteq A^{s+r}$. This and simple induction over s show that for any σ, $[a_1, a_2, \ldots, a_s]_\sigma \in A^s$, which immediately gives the first implication (i) $\Rightarrow$ (ii). Since, obviously, (ii) $\Rightarrow$ (iii), it suffices to show that (iii) $\Rightarrow$ (i).

An easy induction over l shows that, given a set $g_1, g_2, \ldots, g_r, \ldots$ of generators of A, the l-th term A^l of the lower central series of A is generated as an algebra by a set of elements of the form $[g_{i_1}, g_{i_2}, \ldots, g_{i_s}]$, $s \geqq l$. This is enough to prove (iii) $\Rightarrow$ (i). (Compare also with proposition 8.1 below.)

Nilpotent Lie algebras enjoy many 'good' properties. To prove one of them we need a concept of a "*normalizer of a subalgebra B in a Lie algebra A*", denoted by $N_A(B)$. This latter is defined as the set of all $x \in A$ such that $[x, B] \subseteq B$. One immediately observes that $N_A(B)$ is a subalgebra containing B as an ideal.

Proposition 3. *Let A be a nilpotent Lie algebra over some ring k, B be its proper subalgebra. Then* $N_A(B) \neq B$.

Proof. Induction by the nilpotency class c of A, the proposition being trivial for $c = 1$. So, let $c > 1$ and assume, firstly, that $B \not\supseteq A^c$. But then any $x \in A^c \setminus B$ belongs to $N_A(B)$, since $[x, B] = \{0\}$. If, secondly, $B \supseteq L^c$, then B/L^c is a proper subalgebra of A/L^c, the latter algebra being nilpotent of class $c - 1$. Thus the induction applies and $N_{A/L^c}(B/L^c) \neq B/L^c$. This implies by an obvious argument $N_A(B) \neq B$.

Corollary. *Let B be a maximal subalgebra of a nilpotent Lie algebra A. Then B is an ideal of A, A/B being a simple k-module.*

Remark. If A is a nilpotent Lie algebra of class c, B is its subalgebra, A/C is its quotient algebra by some ideal C, then B and A/C are nilpotent of class $\leqq c$. If A possesses an ideal $B \subseteq Z(A)$ such that A/B is nilpotent of class c then A is nilpotent of class $\leqq c + 1$.

All the assertions are immediate corollaries of the above propositions of this section.

1.8. Generators of nilpotent algebras

Let A be a Lie algebra over some ring k, $\{M_\alpha, \alpha \in I\}$ be the (possibly empty) set of its proper maximal subalgebras. The *Frattini subalgebra* $\Phi(A)$ of a Lie algebra A is defined as $\bigcap_{\alpha \in I} M_\alpha$ if $I \neq \emptyset$ or as A if $I = \emptyset$.

Remark. An element $x \in A$ is called a *nongenerator* of A if $A = \text{alg}(x, S)$ for some set S implies $A = \text{alg}(S)$. We assert that $\Phi(A)$ coincides with the set of all nongenerators of A. Indeed, given $x \in A \setminus \Phi(A)$, there exists a maximal proper subalgebra M of A such that $x \notin M$. Then $M = \text{alg}(M) \neq A$, but $A = \text{alg}(x, M)$, and therefore x is not a nongenerator. Conversely, given an element $x \in \Phi(A)$, assume that $A = \text{alg}(x, S)$ but $T = \text{alg}(S) \neq A$. Let M be some subalgebra of A maximal among those ones which contain T but do not contain x. Then M is a maximal subalgebra of A, for if $M_1 \supset M$ then $x \in M_1$ so that $M_1 = A$. But $M \supseteq \Phi(A) \ni x$, contradicting our assumption. Thus $T = \text{alg}(S) = A$, and x is a nongenerator.

Remark also that any finite set of nongenerators is itself a nongenerating set.

Proposition 1. *Let A be a nilpotent Lie algebra over some ring k. Then $\Phi(A)$ is an ideal of A which contains A^2.*

Proof. By the remark above it suffices to prove that any element from A^2 is a nongenerator of A. We prove even that the whole of A^2 is a nongenerating set. So far, let S be a subset of A such that $A = \text{alg}(A^2, S)$. We proceed by induction over the nilpotency class c of A, the case $c = 1$ being trivial. Given, further, $c > 1$, by the induction hypothesis $A/A^c = \text{alg}(S + A^c)$, which yields $A = \text{alg}(S, A^c)$. By the remark from the final paragraph of the proof of proposition 7.2, A^c is spanned by the elements $[g_1, g_2, \dots, g_c]$, g_i being an element from some fixed generating set of A, for example $\{S, A^c\}$. Since all elements from A^c are central, the nonzero commutators spanning A^c have form $[s_1, s_2, \dots, s_c]$, $s_i \in S$, $i = 1, 2, \dots, c$. Therefore, $A^c \subseteq \text{alg}(S)$ and thus $A = \text{alg}(S)$, that is A^2 is nongenerating and $\Phi(A) \supseteq A^2$. The first assertion of the proposition follows from the fact that any subalgebra containing the derived algebra is an ideal.

1.9. Engel's theorem

Recall that a chain

$$0 = M_0 \subseteq M_1 \subseteq \dots \subseteq M_s = M$$

of submodules of a k-module M is called a *composition series* of M if each subsequent factor M_{i+1}/M_i, $i = 0, \dots, s - 1$, of this series is a simple nonzero module. A module M possessing a composition series of finite length s is called a *module of finite length.*

By the Jordan-Hölder theorem any series of a module of finite length can be refined to a composition series, and all composition series have the same length. Therefore, if B is a proper subalgebra of a Lie algebra A having finite length $l(A)$ as a k-module, then $l(B) < l(A)$; if B is a nonzero ideal of A, then $l(A/B) < l(A)$. (Compare with section 15 of this chapter.)

In this section we shall deal with Lie algebras of endomorphisms of k-modulus. Namely, let M be a k-module, $E = \mathrm{End}_k M$ be the Lie algebra of all k-endomorphisms of M.

Lemma. *Let $x \in E$ be an endomorphism such that $x^n = 0$. Then $(\mathrm{ad}\,(x))^{2n-1} = 0$.*

Proof. Given an element $y \in E$, one has

$$(\mathrm{ad}(x))^{2n-1}\,(y) = \sum_{i=0}^{2n-1} \beta_i x^i y x^{2n-1-i} = 0\,, \qquad \beta_i \in k\,,$$

since for any $0 \leqq i \leqq 2n - 1$ either $i \geqq n$ or $2n - 1 - i \geqq n$, that is, either $x^i = 0$ or $x^{2n-1-i} = 0$. This proves the lemma.

Theorem 4. *Given a Lie algebra A of nilpotent endomorphisms of a k-module M, with $l(A) < \infty$, then either $M = \{0\}$, or there exists a nonzero $m \in M$ such that $x(m) = 0$ for all $x \in A$.*

Proof. Use induction over the length of A which we denote by $l(A)$, and assume that $M \neq \{0\}$. First show that if B is some proper subalgebra of A then $N_A(B) \neq B$. Consider the map $\pi\colon B \to \mathrm{End}_k(A/B)$ given by the rule $(\pi(b))\,(a + B) = [b, a] + B$.

The linear map $\pi(b)$ is defined unambiguously since if $a + B = a' + B$ then $a - a' \in B$ and $(\pi(b))\,(a - a') = [b, a - a'] \in B$ for B is a subalgebra. Show that π is a homomorphism of Lie algebras, π being obviously linear. Indeed, $\pi([b_1, b_2])\,(a + B) = [[b_1, b_2], a] + B = [b_1, [b_2, a]] - [b_2, [b_1, a]] + B = (\pi(b_1)\,\pi(b_2) - \pi(b_2)\,\pi(b_1))\,(a) + B$. Moreover, if $b^n = 0$ then by the lemma $(\pi(b))^{2n-1} = 0$. Therefore $\pi(B)$ is a Lie subalgebra of nilpotent endomorphisms of A/B and $l(\pi(B)) \leqq l(B) < l(A)$, so that the induction hypothesis applies. Since A/B is a nonzero module, there exists a nonzero element $a + B \neq B$ such that $\pi(b)\,(a + B) = B$, or $[b, a] \in B$, for all $b \in B$. An element a belongs to $N_A(B) \setminus B$ and we are done.

The just proved property of A shows that any proper maximal subalgebra of A is an ideal. Let now B be some maximal subalgebra of A and $x \notin B$. Then $kx + B = A$. Since $l(B) < l(A)$, a submodule N of all elements $v \in M$ such that $b(v) = 0$ for all $b \in B$ is nonzero. We show that N is A-invariant, that is, $x(N) \subseteq N$. In fact, $b(x(v)) = x(b(v)) + ([b, x])\,(v) = 0$, since $[b, x] \in B$. Element x being nilpotent, there exists a nonzero element $u \in N$ such that $x(u) = 0$. Now obviously u is the desired element and the proof is complete.

Corollary (Engel). *Let A be a Lie algebra of finite length over k and $(\mathrm{ad}\,x)^{n(x)} = 0$ for each $x \in A$. Then A is nilpotent.*

Proof. Consider the Lie algebra ad (A) of the nilpotent endomorphisms of A (see theorem 2). By the just proved theorem there exists an $a \neq 0$ in A such that $(\mathrm{ad}(x))(a) = 0$ for all $x \in A$. This means that $Z(A)$, the centre of A, is nonzero. Consider now $A/Z(A)$; since $l(A/Z(A)) < l(A)$, by induction on $l(A)$ one sees that for some c, $A/Z(A) = Z_{c-1}(A/Z(A))$. Therefore $A = Z_c(A)$ and the corollary follows.

1.10. Two criteria for the nilpotency

Let A, B be Lie algebras over some ring k, C be an ideal of A. We say that A is an *extension of B by C* if $A/C \cong B$. The example of a two-dimensional nonabelian algebra shows that an extension of an abelian Lie algebra by an abelian one may be nonnilpotent. In this section we consider some particular cases when this is not the case.

Theorem 5 (Barnes). *Let A be a Lie algebra of finite length over some ring k, $B \subseteq C$ be ideals of A such that $B \subseteq \Phi(A)$ and C/B is nilpotent. Then C is nilpotent.*

Proof. By our assumptions, there exists some nonnegative integer c such that $C^{c+1} \subseteq B$. We choose then an arbitrary $x \in C$ and show that a k-endomorphism $\varphi = \mathrm{ad}(x)$ is nilpotent.

Now since A has finite length and the chain $\mathrm{Im}\,\varphi \supseteq \mathrm{Im}\,\varphi^2 \supseteq \ldots$ is monotonously nonincreasing there exists a positive integer $t > c$ such that $\mathrm{Im}\,\varphi^{t+s} = \mathrm{Im}\,\varphi^t$ for all $s \geqq 0$. By the same reason $\mathrm{Ker}\,\varphi^{t+s} = \mathrm{Ker}\,\varphi^t$ (obviously, t may be chosen the same in both equations). Given $y \in A$, $\varphi^t(y) = \varphi^{2t}(z)$, for some $z \in A$, so that $u = y - \varphi^t(z) \in \mathrm{Ker}\,\varphi^t$ and, therefore $y \in \mathrm{Ker}\,\varphi^t + \mathrm{Im}\,\varphi^t$. Thus $A = \mathrm{Ker}\,\varphi^t + \mathrm{Im}\,\varphi^t$. Since $t > c$, $\mathrm{Im}\,\varphi^t \subseteq (\mathrm{ad}(x))^{c+1}(A) \subseteq (\mathrm{ad}(x))^c(C) \subseteq C^{c+1} \subseteq \Phi(A)$, and $A = \mathrm{Ker}\,\varphi^t + \Phi(A)$. But $\mathrm{Ker}\,\varphi^t$ is a subalgebra: given $u, v \in \mathrm{Ker}\,\varphi^t$, one has $\varphi^{2t}([u, v]) = \sum_{i=0}^{2t} \beta_i[\varphi^i(u), \varphi^{2t-i}(v)] = 0$, $\beta_i \in k$. Since $\mathrm{Ker}\,\varphi^{2t} = \mathrm{Ker}\,\varphi^t$, $[u, v] \in \mathrm{Ker}\,\varphi^t$, and by the remark from the section 5 (having in mind that $\varphi(A)$ is finitely generated even as k-module), $A = \mathrm{Ker}\,\varphi^t$, so that $\mathrm{ad}(x)$ (acting on C) for any $x \in C$ is nilpotent. By the corollary from the preceding section C is nilpotent, and the proof is complete.

Corollary 1. *Let A be a Lie algebra of finite length over some ring k, $\varphi(A)$ be the maximal ideal of A contained in $\Phi(A)$. Then $\varphi(A)$ is nilpotent.*

The ideal from this corollary is called the *Frattini ideal* of A.

The proof of the next corollary is left to the reader, although the more general result is proved some few lines below.

Corollary 2. *Let A be a Lie algebra of finite length over some ring k, B be its nilpotent ideal such that L/B^2 is nilpotent. Then A is nilpotent.*

Now the promised more general result, supplying a criterion for nilpotence without restrictions on the length.

Theorem 6. *Given a Lie algebra A over some ring k, having a nilpotent ideal B with nilpotent A/B^2. Then A is nilpotent.*

Proof. Induction by the nilpotency class c of B, the assertion being trivial if $c = 1$. The subalgebra B^c is easily seen to be an ideal of A and thus the induction hypothesis applies to A/B^c. Assuming A/B^c nilpotent of class, say d, any element $a \in B^c$ has the form $a = \sum_i [m_i, n_i]$, $m_i, n_i \in B$. We calculate then $[x_{2d-1}, \ldots, x_2, x_1, a]$. By the definition of the derivation, we have

$$[x_{2d-1}, \ldots, x_2, x_1, a] = \sum_i \sum [[x_{i_1}, \ldots, x_{i_s}, m_i], [x_{j_1}, \ldots, x_{j_{2d-1-s}}, n_i]]$$

By a standard argument in each summand one of the factors lies in B^c (equals zero in A/B^c), and since the remaining factor lies in B, the whole summand equals zero and thus A is nilpotent of class $\leq 3d - 1$.

The theorem is proved.

1.11. Solvable Lie algebras

Call a Lie algebra A over some ring k *solvable of (solvable) length* l if $A^{(l-1)} \neq \{0\}$ but $A^{(l)} = \{0\}$ (cf. section 6 for the definition of the derived series). Thus, for example, any abelian (nonzero) algebra is solvable of length l.

Proposition 1. *Let A possess a finite series*

$$A = A_0 \supseteq A_1 \supseteq \dots \supseteq A_s = \{0\}$$

of subalgebras such that A_{i+1} is an ideal of A_i, $i = 0, 1, \dots, s-1$, and A_i/A_{i+1} is abelian. Then A is solvable of length at most s. (The converse is obviously true.)

Proof. The only thing to be proved is an easy statement: given an algebra S with an ideal T such that S/T is abelian, then $T \supseteq S'$. This and an evident inductive argument completing the proof is left to the reader.

Examples. (i) By the preceding proposition every nilpotent Lie algebra is solvable.

(ii) Any extension of a solvable Lie algebra by a solvable one is solvable. Solvable algebras of length 2 are called *metabelian*: they are extensions of abelian algebras by abelian ones.

(iii) The next example is typical for Lie algebras over an algebraically closed field of characteristic 0. Let A be a Lie algebra of (say, upper) triangular m by m matrices over some ring k, that is

$$A = \left\{ \begin{pmatrix} a_{11} & a_{12} & \dots & a_{1m} \\ 0 & a_{22} & \dots & a_{2m} \\ \dots & \dots & \dots & \dots \\ 0 & 0 & \dots & a_{mm} \end{pmatrix}, \ a_{ij} \in k \right\}.$$

Then A' is a subalgebra of the algebra $nt_m(k)$ of niltriangular m by m matrices which is easily seen to be nilpotent of class $m - 1$. Thus by (i) and (iii) A is solvable. A is denoted by $t_m(k)$.

(iv) Any two-dimensional algebra is metabelian.

(v) Any subalgebra and any quotient algebra of a solvable Lie algebra is solvable.

1.12. Higgins' theorem

We start with an

Example. Let k be a commutative ring with unity of prime additive exponent p, M be a free k-module with free basis $\{e_f, f \in \mathbf{Z}_p^{\mathbf{N}}\}$ where $\mathbf{Z}_p^{\mathbf{N}}$ is the set of all $\mathbf{Z}_p$-valued functions of natural argument. Let further δ_m be the endomorphism of M given by $\delta_m(e_f) = e_{\bar{f}}$, where $\bar{f}(i) = f(i)$ if $i \neq m$, and $\bar{f}(m) = f(m) + 1$; ε be the identity map of M. Now given an abelian Lie algebra A, which is a free k-module with free basis $\{x_i, i \in \mathsf{N}\}$, consider the isomorphism $\varphi: A \to \operatorname{End}_k(M)$ extending the map $\varphi: x_i \mapsto \delta_i - \varepsilon$, $i = 1, 2, \ldots$, M being considered as an abelian Lie algebra, form a semidirect product $T = A \curlywedge M$ of A by M subject to φ. Since $(\delta_i - \varepsilon)^p = \delta_i^p - \varepsilon^p = 0$, the definition of a semidirect product implies that $(\operatorname{ad}(x_i))^p (e_f) = 0$, for all $i \in \mathsf{N}$, $f_m \in \mathsf{Z}_p^{\mathsf{N}}$. Given then $t, u \in T$ such that $t = \sum_s \alpha_s x_s + \sum_f \beta_f e_f$, one has $[t, u] \in T$ and

$$\big(\operatorname{ad}(t)\big)^{p+1} (u) = \big(\operatorname{ad}(t)\big)^p ([t, u]) = \Big(\operatorname{ad}\big(\sum_s \alpha_s x_s\big)\Big)^p ([t, u]) =$$

$$= \sum_s \alpha_s^p (\operatorname{ad}(x_s))^p ([t, u]) = 0 .$$

Thus $(\operatorname{ad}(t))^{p+1} = 0$ for all $t \in T$ and hence T, being a Lie algebra of infinite length over k, satisfies the other condition of the corollary from the section 9. However T is nonnilpotent, for if $f \in \mathsf{Z}_p^{\mathsf{N}}$ is given by $f_m(i) = 1$, $l \leqq i \leqq m$, and $f_m(i) = 0$, $i > m$, then

$$[x_m, x_{m-1}, \ldots, x_1, e_0] = e_{f_m} \neq 0 .$$

T being metabelian, the example justifies the restrictions both in Engel's theorem and in a theorem due to Higgins which follows.

Theorem 7. *Let A be a Lie algebra over some ring k with unity 1, n be a positive integer such that $(n!)\,1$ is invertible in k. Assume that A is solvable of length l and that $\big(\operatorname{ad}(x)\big)^n = 0$ for any $x \in A$. Then A is nilpotent of class*

$$c \leqq \frac{n^l - 1}{n - 1} .$$

We need two auxiliary lemmas.

Lemma 1. *Let C be a subgroup of the additive group of an associative algebra B over some ring k such that for some positive integer n, $v^n = 0$ for all $v \in C$. Then for all $v_1, v_2, \ldots, v_n \in C$,*

$$\sum_{\pi \in S_n} v_{\pi(1)} v_{\pi(2)} \ldots v_{\pi(n)} = 0 , \tag{1}$$

where S_n is the symmetrical group on symbols $1, 2, \ldots, n$.

Proof. Denote the left hand side of (1) by w. Given a set $\lambda_1, \lambda_2, \ldots, \lambda_n$ of indeterminates, consider an expression

$$a = (\lambda_1 v_1 + \lambda_2 v_2 + \ldots + \lambda_n v_n)^n . \tag{2}$$

By the condition, for integers $\lambda_1, \lambda_2, \ldots, \lambda_n$ this expression equals zero. Rewrite then a in the form

$$a = r_0 + \lambda_1 r_1 + \lambda_1 \lambda_2 r_2 + \ldots + \lambda_1 \lambda_2 \ldots \lambda_n w , \tag{3}$$

where r_0 is the sum of those monomials from (2) which are not divisible by λ_1, $\lambda_1 r_1$ is the sum of those monomials from $a - r_0$ which are not divisible by λ_2, and so on. Putting in (3) firstly $\lambda_1 = 0$, we obtain $r_0 = 0$. Putting then in (3) $\lambda_1 = 1$, $\lambda_2 = 0$, we obtain $r_1 = 0$, and so on. Finally, putting $\lambda_n = 1$, we obtain $w = 0$, and the proof is complete.

In particular, let $B = \operatorname{End}_k(A)$, $C = \operatorname{ad}(A)$. Then by the just proved lemma for all $x_1, x_2, \dots, x_n \in A$,

$$\sum_{\pi \in S_n} \big(\operatorname{ad}(x_{\pi(1)})\big)\big(\operatorname{ad}(x_{\pi(2)})\big) \dots \big(\operatorname{ad}(x_{\pi(n)})\big) = 0 .$$

Lemma 2. *Under the conditions of the theorem 7, let $M = A'$, M^i, $i = 1, 2, \dots$, be members of the lower central series of M. Then*

$$\big(\operatorname{ad}(x_{\pi(1)})\big)\big(\operatorname{ad}(x_{\pi(2)})\big) \dots \big(\operatorname{ad}(x_{\pi(n)})\big)(M^i) + M^{i+1} =$$
$$= \big(\operatorname{ad}(x_1)\big)\big(\operatorname{ad}(x_2)\big) \dots \big(\operatorname{ad}(x_n)\big)(M^i) + M^{i+1} ,$$

for all $x_1, x_2, \dots, x_n \in A$ and $i = 1, 2, \dots$.

Proof. We proceed by induction on the number of inversions in the substitution $\pi \in S_n$, the case when this number is 0 being trivial. So let r, $1 \leqq r < n$, be such an integer that $\pi(r) > \pi(r+1)$. Then

$$\big(\operatorname{ad}(x_{\pi(1)})\big)\big(\operatorname{ad}(x_{\pi(2)})\big) \dots \big(\operatorname{ad}(x_{\pi(r)})\big)\big(\operatorname{ad}(x_{\pi(r+1)})\big) \dots \big(\operatorname{ad}(x_{\pi(n)})\big) =$$
$$= \big(\operatorname{ad}(x_{\pi(1)})\big) \dots \big(\operatorname{ad}(x_{\pi(r+1)})\big)\big(\operatorname{ad}(x_{\pi(r)})\big) \dots \big(\operatorname{ad}(x_{\pi(n)})\big) +$$
$$+ \big(\operatorname{ad}(x_{\pi(n)})\big) \dots \Big(\Big(\big(\operatorname{ad}(x_{\pi(r)})\big)\big(\operatorname{ad}(x_{\pi(r+1)})\big) - \big(\operatorname{ad}(x_{\pi(r+1)})\big)\big(\operatorname{ad}(x_{\pi(r)})\big)\Big)\Big) \dots \big(\operatorname{ad}(x_{\pi(n)})\big) =$$
$$= \big(\operatorname{ad}(x_{\pi'(1)})\big) \dots \big(\operatorname{ad}(x_{\pi'(r)})\big)\big(\operatorname{ad}(x_{\pi'(r+1)})\big) \dots \big(\operatorname{ad}(x_{\pi'(n)})\big) +$$
$$+ \big(\operatorname{ad}(x_{\pi(1)})\big) \dots \big(\operatorname{ad}([x_{\pi(r)}, x_{\pi(r+1)}])\big) \dots \big(\operatorname{ad}(x_{\pi(n)})\big) .$$

The first summand can be dealt with by the induction hypothesis, and

$$\big(\operatorname{ad}(x_{\pi(1)})\big) \dots \big(\operatorname{ad}([x_{\pi(r)}, x_{\pi(r+1)}])\big) \dots \big(\operatorname{ad}(x_{\pi(n)})\big)(M^i) \subseteq M^{i+1}$$

for M^i and M^{i+1} are ideals of A and $[x_{\pi(r)}, x_{\pi(r+1)}] \in M$.

Thus the lemma is proved.

Proof of the theorem. Induction by the solvable length l of A, the case $l = 1$ being trivial. Since $M = L'$ has solvable length $l - 1$, the induction hypothesis applies and therefore M is nilpotent of class $c' \leqq \dfrac{n^{l-1} - 1}{n - 1}$. Using lemmas 1 and 2 we get

$$\big(\operatorname{ad}(x_1)\big) \dots \big(\operatorname{ad}(x_n)\big)(M^i) + M^{i+1} =$$
$$= \frac{1}{(n!)\,1} \sum_{\pi \in S_n} \big(\operatorname{ad}(x_{\pi(1)})\big) \dots \big(\operatorname{ad}(x_{\pi(n)})\big)(M^i) + M^{i+1} = M^{i+1} , \qquad i = 1, 2, \dots .$$

Since $\big(\operatorname{ad}(x_0)\big)(A) \subseteq M$ and $M^{c'+1} = \{0\}$, for all $x_0, x_1, \dots, x_{c'n}$ from A one has

$$\big(\operatorname{ad}(x_{c'n})\big) \dots \big(\operatorname{ad}(x_1)\big)\big(\operatorname{ad}(x_0)\big) = 0 .$$

Then the nilpotency class of A does not exceed $c'n + 1 = \dfrac{n^l - 1}{n - 1}$, and the proof is complete.

1.13. The Frattini subalgebra

In this section we again turn our concern to Lie algebras having finite length over the base ring k. But first one more definition. Given a set S of elements in a Lie algebra A, the set $C_A(S)$ of all elements $x \in A$ with $[x, S] = \{0\}$ is called a *centralizer* of S in A.

Remark that if S is an abelian ideal of A then $C_A(S)$ is an ideal of A (containing S). Remark also, that if S in any minimal ideal of a solvable Lie algebra A, then S is abelian. Indeed, since S is solvable, the last nonzero term $S^{(l-1)}$ of the derived series of S is an abelian ideal of A contained in S. Therefore $S^{(l-1)} = S$, and S is abelian. Now our first theorem about Frattini subalgebras in the nonnilpotent case.

Theorem 8 (Barnes). *Let k be a commutative ring, A be a solvable Lie algebra of finite length as a k-module, $\Phi(A)$ be a Frattini subalgebra of A. Then $\Phi(A)$ is an ideal of A.*

Proof. We proceed by induction over the length $l(A)$ of A as a k-module, the case $l(A) = 1$ being trivial. Let $l(A) > 1$ and fix a minimal ideal M of A. Denote by Φ_M the intersection of all maximal subalgebras of A containing M. Since, clearly, $\Phi_M/M = \Phi(A/M)$, Φ_M is an ideal of A by the induction hypothesis. Thus if $M \subseteq \Phi(A)$ then $\Phi(A) = \Phi_M$ is an ideal. Assume therefore that $M \nsubseteq \Phi(A)$, so that $A = M \oplus V$, V being an arbitrary maximal proper subalgebra of A with $V \nsupseteq M$, and the sum being direct since $M \cap V$ is an ideal in A. Suppose now that $C_A(M) \supsetneqq M$. Then $V \cap C_A(M)$ is a nonzero ideal of A contained in V. Therefore, any maximal subalgebra contains a minimal ideal of A and hence $\Phi(A)$ is an intersection of Φ_M, M taken from the set of all minimal ideals of A. This argument proving $\Phi(A)$ to be an ideal of A, it suffices to consider the case $C_A(M) = M$. We show that in this case $\Phi(A) = \{0\}$. Suppose not, and let x be a nonzero element from $\Phi(A)$. Choose a maximal subalgebra $V \nsupseteq M$. Then $x \in V \setminus M$, and there exists an $m \in M$ such that $[x, m] \neq 0$. Since $(\mathrm{ad}(m))^2 = 0$, the linear map $1 + \mathrm{ad}(m)$ is easily seen to be an automorphism of A (cf. 3,8). Let $x' = (1 + \mathrm{ad}(m))\,(x) = x + [m, x]$ be the image of x under this automorphism. Then, since $\Phi(A)$ obviously admits all the automorphisms of A, $x' \in \Phi(A)$, and thus $x' \in V$. On the other hand one has for any $m' \in M$, $[m', x'] = [m', x] + [m', [m, x]] = [m', x]$ and $[m', x' - x] = 0$. By our assumptions then $x' - x \in C_A(M) \cap V = \{0\}$, that is $x = x'$. But then $[m, x] = 0$ contrary to the choice of m. The theorem is proved.

In structure theory of Lie algebras of finite length two ideals, termed as *solvable* and *nilpotent radicals*, play an important role. They are defined in the following

Proposition 1. *Let A be a Lie algebra of finite length over some ring k. Then the sum S of all solvable ideals of A is the greatest solvable ideal of A, called the solvable radical of A and denoted by $S(A)$. The sum N of all nilpotent ideals of A is the greatest nilpotent ideal of A, called the nilpotent radical of A and denoted by $N(A)$. Moreover, $S(A/S(A)) = \{0\}$.*

Proof. The only thing worth proving is the fact that the sum of two nilpotent ideals U, V is again a nilpotent ideal. So, let $U^s = V^r = \{0\}$. Prove that $(U + V)^{s+r+1} = \{0\}$. Indeed, $(U + V)^{s+r+1}$ is generated by the commutators of the form $[w_{s+r+1}, w_{s+r}, \dots, w_1]$, where w_i is an element either from U or from V. Assume that $w_{i_1}, w_{i_2}, \dots, w_{i_t} \in U$ and $t \geqq s + 1$. Then, since U^i is an ideal of $U + V, i = 1, 2, \dots, s$, one has inductively, $[w_{i_1}, \dots, w_1] \in U$, $[w_{i_2}, \dots, w_{i_1}, \dots, w_1] \in U^2, \dots, [w_{i_t}, \dots, w_1] \in U^t = \{0\}$. If $t < s + 1$, then the same argument applies to factors from V whose number is $\geqq r + 1$. Thus the proposition is proved.

Let A be a Lie algebra. By $S_0(A)$ we denote the sum of all minimal abelian ideals of A. This ideal, called the *abelian socle* of A, is easily seen to be equal to the maximal direct sum of some of these ideals. Our goal in the remaining part of this section, is the following

Theorem 9 (Towers). *Let A be a Lie algebra of finite length over k, $\Phi(A) = \Phi$, $\varphi(A) = \varphi$, $N(A) = N$. Then $\varphi(A/\varphi) = \{0\}$ and*

(i) $\Phi \cap N = \varphi \supseteq N^2$;

(ii) $N(A/\varphi) = N/\varphi = S_0(A/\varphi)$;

(iii) *N/φ is complemented in A/φ, that is, there exists a subalgebra M/φ such that $A/\varphi = N/\varphi \oplus M/\varphi$.*

Proof. The fact $\varphi(A/\varphi) = \{0\}$ being trivial (for the definition of the Frattini ideal see section 10), show (i).

Lemma 1. *Let C be an ideal of A contained in $\Phi(B)$, B being a subalgebra of A. Then $C \subseteq \Phi(A)$.*

Suppose not. Then for some maximal subalgebra M of A one has $C + M = A$. Using the modular law, write $C + M \cap B = A \cap B = B$. Since $C \subseteq \Phi(B)$, $M \cap B = B$, which is a contradiction since then $M \supseteq B \supseteq C$. Thus the lemma is proved.

Now the inclusion $\Phi(A) \cap N \supseteq \varphi(A)$ being evident by the theorem 5, assume that $\Phi(A) \cap N \nsubseteq \varphi(A)$. Since N is nilpotent, $\Phi(N) \supseteq N^2$, and thus, by lemma 1, $\Phi(A) \supseteq N^2$. Put $\Phi_1 = \Phi(A) \cap N$. If $\Phi_1 \lhd A$ then $\varphi \supseteq \Phi_1$, a contradiction; so $\Phi_1 \ntriangleleft A$ and hence $[\Phi_1, A] \nsubseteq \Phi$. Let M be a maximal subalgebra of A with $[\Phi_1, A] \nsubseteq M$. Then $A = N + M$, and multiplying both sides of this equation by Φ_1 we have: $[A, \Phi_1] = [\Phi_1, N + M] \subseteq N^2 + M \subseteq \Phi(A) + M \subseteq M$, a contradiction to the choice of M. So $\Phi \cap N = \varphi \supseteq N^2$ and (i) is proved.

Consider then (ii) and (iii)

Lemma 2. *Let A be a Lie algebra with $\varphi(A) = \{0\}$, B be an abelian ideal of A. Then B is complemented in A.*

Indeed, let T be a subalgebra of A minimal among those ones for which $T + B = A$. Consider $B \cap T$. If $B \cap T \subseteq \varphi(T)$ then by lemma 1 $B \cap T$ (being an ideal of A) $\subseteq \varphi(A) = \{0\}$ and we are done. Otherwise T possesses a proper maximal subalgebra M such that $T = M + (B \cap T)$. But in this case $A = B + (B \cap T + M) = B + M$ contradicting the choice of T.

Clearly (iii) immediately follows from this lemma. The first equality in (ii) is implied by the theorem 5: evidently $N/\varphi \subseteq N(A/\varphi)$; if further T/φ is a nilpotent ideal of A then T is a nilpotent ideal of A so that $T \subseteq N$. As well simple is the second equation. For, clearly, $S_0 = S_0(A/\varphi) \subseteq N(A/\varphi)$. Let then M/φ be a complement to S_0 in A/φ. Then $A/\varphi = S_0 \oplus M/\varphi$ and hence $N/\varphi = S_0 \oplus (M/\varphi \cap N/\varphi)$. Since N/φ by (i) is abelian, $M/\varphi \cap N/\varphi$ is an abelian ideal of A/φ. If B/φ is a minimal nonzero ideal of A/φ contained in $M/\varphi \cap N/\varphi$, then $B/\varphi \subseteq S_0$ contradicting the directness of the above written decompositions. Therefore $N/\varphi = S_0/\varphi$ and (ii) is complete.

Thus the theorem is proved.

1.14. Restricted Lie algebras

In this section we introduce one more class of universal algebras also arising from the consideration of associative algebras. The base ring k will be a commutative ring with unity of prime characteristic (additive exponent) p. So far, let R be an associative algebra over k, the multiplication in R being denoted by juxtaposition. Consider R as a universal algebra with respect to two derived operations : (1) $(x, y) \mapsto [x, y] = xy - yx$; (ii) $x \to x^p$

Proposition 1. *The following identical relations hold in* R:

(a) $(\lambda x)^p = \lambda^p x^p, \lambda \in k, x \in R$;

(b) $(\mathrm{ad}(x))^p = \mathrm{ad}(x^p), x \in R$;

(c) $(x + y)^p = x^p + y^p + \sum_{i=1}^{p-1} s_i(x, y)$, *$is_i(x, y)$ is the coefficient at λ^{i-1} in the polynomial* $(\mathrm{ad}(\lambda x + y))^{p-1}(x) \in R[\lambda]$, $x, y \in R$.

Proof. (a) being trivial, we prove (b). Put $l_x(y) = xy, r_x(y) = yx$; clearly $l_u r_v = r_v l_u$. Then $\mathrm{ad}(x) = l_x - r_x$ and $(\mathrm{ad}(x))^p = (l_x - r_x)^p = \sum_{i=0}^{p} (-1)^i \binom{p}{i} l_x^i r_x^{p-i} = l_x^p - r_x^p = l_{x^p} - r_{x^p} = \mathrm{ad}(x^p)$, the most part of members of the sum being equal zero by the choice of k and $(-1)^p r_x^p = -r_x^p$ since p is a prime number. Thus (b) is complete.

We prove (c). Consider in $R[\lambda]$ the following equation

$$(\lambda x + y)^p = \lambda^p x^p + y^p + \sum_{i=1}^{p-1} \lambda^i s_i(x, y). \tag{1}$$

Deriving both parts of (1) by λ one obtains

$$\sum_{i=0}^{p-1} (\lambda x + y)^i x (\lambda x + y)^{p-i-1} = \sum_{i=1}^{p-1} i\lambda^{i-1} s_i(x, y)\,. \tag{2}$$

Just as in the proof of (b) one deduces that the left hand side of (2) equals to $(\mathrm{ad}(\lambda x + y))^{p-1}(x)$, this however proving (c).

Remark here that by (c) each $is_i(x, y)$ is a "Lie element in x, y", that is $is_i(x, y)$ is an element of a Lie subalgebra of R generated by x, y. Since $1 \leqq i \leqq p - 1$, p being prime, this yields that $s_i(x, y)$ is itself an element of a Lie subalgebra $\mathrm{alg}(x, y)$ of R.

Using this proposition we introduce a class of "*restricted* Lie algebras" or "*Lie p-algebras*" as a class of Lie algebras over some ring k of prime characteristic p endowed with an additional unary operation denoted by $x \mapsto x^p$ and connected with other operations by axioms (a), (b), (c).

A *p-subalgebra* (*p-ideal*) of a Lie p-algebra A is an ordinary Lie subalgebra (ideal) B of A such that given an element $x \in B$, one has $x^p \in B$. A "*p-homomorphism*" of a Lie p-algebra A into a Lie p-algebra B is an ordinary Lie homomorphism $\varphi: A \to B$ such that $\varphi(x^p) = (\varphi(x))^p$ for each $x \in A$. Adding prefix "p-" in suitable places one reformulates and easily proves theorem 1, for Lie p-algebras.

A proposition which follows links ordinary and p-subalgebras in Lie p-algebras, the least p-subalgebra containing subset X being denoted by $\mathrm{alg}_p(X)$.

Proposition 2. *Given a Lie p-algebra A and an (ordinary) subalgebra B spanned as a k-module by a set X of elements of A. Then the p-subalgebra $C = \mathrm{alg}_p(B)$ generated by B is spanned as a k-module by a set Y of elements* x^{p^r} $(= (x^{p^{r-1}})^p)$, $x \in X$, $r = 0, 1, 2, \ldots$.

Proof. It suffices to verify that the set of elements of the form

$$\alpha_1 x_1^{p^{r_1}} + \alpha_2 x_2^{p^{r_2}} + \ldots + \alpha_s x_s^{p^{r_s}}, \qquad \alpha_1, \alpha_2, \ldots, \alpha_s \in k; \qquad x_1, x_2, \ldots, x_s \in X;$$

$r_1, r_2, \ldots, r_s$ being nonnegative integers, forms a p-subalgebra of A. Indeed, given $x, y \in X$, $[x^{p^u}, y^{p^v}] = (\mathrm{ad}(x^{p^u}))(y^{p^v}) = (\mathrm{ad}(x))^{p^u}(y^{p^v}) = (\mathrm{ad}(x))^{p^u-1}([x, y^{p^v}]) =$

$= -(\mathrm{ad}(x))^{p^u-1}(\mathrm{ad}(y))^{p^v}(x) \in \mathrm{alg}(\{x, y\}) \subseteq H$. Using this remark and axiom (c) one easily observes that

$$(\alpha_1 x_1^{p^{r_1}} + \alpha_2 x_2^{p^{r_2}} + \dots + \alpha_s x_s^{p^{r_s}})^p = \alpha_1^p x_1^{p^{r_1+1}} + \alpha_2 x_2^{p^{r_2+1}} + \dots + \alpha_s x_s^{p^{r_s+1}} + h\,,$$

where $h \in H$. This proves our proposition.

Examples. (i) Given some algebra A, $\mathrm{Der}_k(A)$ is a p-subalgebra of $\mathrm{End}_k(A)$, that is, if $d \in \mathrm{Der}_k(A)$ then $d^p \in \mathrm{Der}_k(A)$, k being a ring of characteristic $p > 0$. In particular, let $A = k[x]/(x^p)$, $B = \mathrm{Der}_k(A)$. Then B is spanned by derivations d_i, $i = 0, 1, 2, \dots, p-1$ such that $d_i(x) = x^i$. One sees that $[d_i, d_j] = (j-i)\, d_{i+j-1}$ and therefore $B \cong W_1$, a Lie algebra from the example (ii), section 2. Being isomorphic to the algebra of all derivations of some algebra, W_1 is a p-algebra under the map $d_i \mapsto d_i^p$, $i = 0, 1, 2, \dots, p-1$.

(ii) Let $sl_n(k)$ be the Lie algebra of all matrices of trace 0 over k, $\mathrm{char}(k) = p > 0$. Then $sl_n(k)$ is a p-subalgebra of k_n and thus a restricted Lie algebra. (In fact, $\mathrm{tr}(U) = 0 \Rightarrow \mathrm{tr}(U^p) = 0$.)

(iii) Every Lie algebra over a ring k of characteristic $p > 0$ isomorphically embedded in some associative algebra can be isomorphically embedded in a restricted Lie algebra.

1.15. Modules

Let M be a module over some commutative ring k with unity and A be a Lie algebra over k. We say that M is an *A-module* if there is given a bilinear map $A \times M \to M$ written through $(a, m) \mapsto am$ such that

$$[a_1, a_2]\, m = a_1(a_2 m) - a_2(a_1 m)\,. \tag{1}$$

A k-submodule N is called an *A-submodule* of M if $AN \subseteq N$. If N is some A-submodule then the k-module M/N is canonically an A-module through

$$a(m + N) = am + N\,,$$

the definition being correct by the closure property of N. We call M/N a *quotient module* of M by N.

Examples. (i) Given a k-module M and a Lie k-algebra A, a homomorphism of Lie algebras $\varphi: A \to \mathrm{End}_k(M)$ is called a *representation* of A by endomorphisms of M. In this case M may be canonically turned into an A-module, if one puts

$$am = (\varphi(a))\,(m)\,.$$

The axiom (1) holds since $\varphi([a_1, a_2]) = [\varphi(a_1), \varphi(a_2)] = \varphi(a_1)\,\varphi(a_2) - \varphi(a_2)\,\varphi(a_1)$ by the definition of bracketing in the associative algebra $\mathrm{End}_k(M)$.

Conversely, if M is an A-module, then $a \mapsto a_M$, where an endomorphism $a_M: M \to M$ is defined by $a_M(m) = am$, is a homomorphism of A into $\mathrm{End}_k(M)$ and thus the concept of a Lie module is in some sense equivalent to that of a representation. In particular if $A \subseteq \mathrm{End}_k(M)$, then M is canonically an A-module.

(ii) Using this remark one may construct numerous examples of Lie modules. For example, any Lie algebra A is an A-module through $ax = [a, x] = (\mathrm{ad}(a))\,(x)$, for

$a \mapsto \mathrm{ad}(a)$ is a homomorphism of A into $\mathrm{End}_k(A)$ by theorem 2, (iii). Its submodules are exactly its ideals. If B is an ideal of A then A/B is also an A-module, the quotient module of the A-module A. Moreover, if $D = C/B$ is a factor of A such that both C and B are ideals of A then D is canonically an A-module. A submodule C/B of the A/B-module A/B is a submodule of the A-module A/B and conversely.

(iii) If H is a subalgebra of A then A is an H-module and H is its submodule (cf. the proof of theorem 4).

(iv) If M is an A-module, $C = C_A(M)$ is its centralizer (or annihilator) in A, i.e. $C = \{a \mid aM = 0\}$, then C is an ideal in A and M is canonically an A/C-module. To define the action of A/C on M denote by $\bar{a}$ the element $a + C \in A/C$ and put $\bar{a}m = am$.

In particular, if B is an abelian ideal then $C \supseteq B$ and thus B may be considered as an A/B-module, the action being defined by the same rule: $(a + B)\, b = [a, b]$.

An A-module M is called an *irreducible* (or *simple*) one if its only A-submodules are $\{0\}$ and M itself. *Fully reducible* (or *semisimple*) is an A-module M which is a (direct) sum of irreducible A-submodules. Important properties of fully reducible modules are given in the following

Proposition 1. *Let N be a submodule of a fully reducible A-module M. Then for some A-submodule N', $M = N \oplus N'$. Besides, N and M/N are fully reducible.*

Proof. Let $M = \bigoplus\limits_{\alpha} M_\alpha$, M_α being irreducible A-submodules of M, and N' be a maximal A-submodule of M with $N \cap N' = \{0\}$. Assume that $N \oplus N' \neq M$, then some M_α, say, M_1, is not a submodule of $N \oplus N'$, and therefore $(N \oplus N') \cap M_1 = \{0\}$. This yields $N \cap (N' \oplus M_1) = \{0\}$ and contradicts our choice of N'. Therefore $N \oplus N' = M$ and the first statement is complete.

To prove the second one denote by Q the sum of all irreducible A-submodules of N and suppose $Q \neq N$. By the first part for some A-submodule Q' of M one has $M = Q \oplus Q'$. Using the modular law, one has

$$N = N \cap M = N \cap (Q \oplus Q') = Q \oplus N \cap Q' .$$

By our assumption, $N \cap Q' \neq \{0\}$, and hence $N \cap Q'$ contains some minimal A-submodule N_1. But then $N_1 \subseteq Q' \cap Q = \{0\}$ which is a contradiction proving our statement about N. The latter statement about M/N evidently follows from two former ones.

Remark that we have not used the fact that A is a Lie algebra. The same is true for one more proposition 2, which we preface by a definition. Given an A-module M, a series

$$\{0\} = M_0 \subset M_1 \subset \dots \subset M_s = M \tag{2}$$

of its A-submodules is called a *chief* one if for each $i = 1, \dots, s$, M_i/M_{i-1} is an irreducible nonzero A-module. In other words each M_{i-1} is a maximal A-submodule of M_i distinct from M_i.

Let

$$\{0\} = N_0 \subset N_1 \subset \dots \subset N_t = M \tag{3}$$

be another possibly not chief series of M. Then

Proposition 2. *The lengths of all chief series of M are the same. Moreover any chief factor N_i/N_{i-1} of (3) is isomorphic to some M_j/M_{j-1} of (2).*

Proof. We introduce submodules $M_{ij} = M_i + (M_{i+1} \cap N_j)$ and $N_{ji} = N_j + N_{j+1} \cap M_i$ and prove that

$$M_{ij}/M_{i,j-1} \cong N_{ji}/N_{j,i-1}\,.$$

Indeed, by the isomorphism theorem

$$\begin{aligned} M_{ij}/M_{i,j-1} &= M_{i-1} + (M_i \cap N_j)/M_{i-1} + M_i \cap N_{j-1} = \\ &= M_{i-1} + M_i \cap N_{j-1} + M_i \cap N_j/M_{i-1} + M_i \cap N_{j-1} \cong \\ &\cong M_i \cap N_j/(M_{i-1} + M_i \cap N_{j-1}) \cap M_i \cap N_j = \\ &= M_i \cap N_j/M_{i-1} \cap N_j + M_i \cap N_{j-1} = \\ &= M_i \cap N_j/(M_{i-1} \cap N_j + N_{j-1}) \cap M_i \cap N_j \cong \\ &\cong M_i \cap N_j + M_{i-1} \cap N_j + N_{j-1}/N_{j-1} + M_{i-1} \cap N_j = N_{j,i}/N_{j,i-1}\,. \end{aligned}$$

Since (2) is a chief series, for each $i = 1, 2, \dots, s-1$, there exists only one j_i such that $M_i/M_{i-1} = M_{i,j_i}/M_{i,j_i-1} \cong N_{j_i,i}/N_{j_i,i-1}$. This gives a one-to-one correspondence between factors of (2) and of the refinement of (3) by submodules N_{ji}. Therefore $s \geqq t$. In particular, if (3) is also chief, $s = t$ and $M_i/M_{i-1} \cong N_{j_i}/N_{j_i-1}$ which completes the proof of the proposition.

A generalization of a chief series is a *fully reducible series*, that is a series (2) in which each factor M_i/M_{i-1} is a fully reducible A-module. The next assertion concerns with such series.

Proposition 3. *Let N be a submodule of some A-module M possessing some fully reducible series* (2) *of the length s. Then M/N also possesses a fully reducible series of some length $\leqq s$.*

Proof. Consider a chain of A-submodules in M/N of the form

$$\begin{aligned} &N/N \subseteq M_1 + N/N \subseteq \dots \subseteq M_{i-1} + N/N \subseteq M_i + N/N \subseteq \dots \\ &\dots \subseteq M + N/N = M/N\,. \end{aligned} \tag{4}$$

Then its factors may be transformed as follows: $M_i + N/N/M_{i-1} + N/N \cong \cong M_i + N/M_{i-1} + N = M_i + (M_{i-1} + N)/M_{i-1} + N \cong M_i/M_i \cap (M_{i-1} + N) = = M_i/M_{i-1} + M_i \cap N \cong (M_i/M_{i-1})/(M_{i-1} + M_i \cap N/M_{i-1})$, that is, they are quotient modules of fully reducible modules M_i/M_{i-1} by their submodules $M_{i-1} + M_i \cap N/M_{i-1}$. The proposition 1 now completes the proof of the present assertion.

To deduce some useful corollaries from these propositions, remark that a chief series of a Lie algebra A considered canonically as an A-module (example (ii)) is called simply a chief series of a Lie algebra A. Clearly in many cases such series do not exist, but if A has finite length over some commutative ring k they certainly exist. Now the promised

Corollary. *Given a Lie algebra A and its chief series*

$$\{0\} = A_0 \subset A_1 \subset \dots \subset A_{i-1} \subset A_i \subset \dots \subset A_n = A\,. \tag{5}$$

Then

(i) *any chief factor P/Q of A is A-isomorphic with some factor A_i/A_{i-1} of the series* (5);

(ii) *any chief factor of some subalgebra $B \subseteq A$ is isomorphic with some B-subfactor of a B-module A_i/A_{i-1} for some $0 < i \leqq n$ (B is assumed to have at least one chief series)*;

(iii) *if H is an ideal of A and* (5) *is a fully reducible series of A then H possesses a fully reducible series having length $\leqq n$.*

Proof. The statement (i) follows from proposition 2 applied to (5) and the A-series $\{0\} \subseteq Q \subset P \subseteq A$. The (ii) one is valid since a chief series for B may be obtained by the refinement of the series of the form

$$\{0\} \subseteq B \cap A_0 \subseteq B \cap A_1 \subseteq \ldots \subseteq B \cap A_{i-1} \subseteq B \cap A_i \subseteq \ldots$$
$$\ldots \subseteq B \cap A_n = B\,, \tag{6}$$

whose factors have the form $B \cap A_i/B \cap A_{i-1} = B \cap A_i/(B \cap A_i) \cap A_{i-1} \cong B \cap A_i + A_{i-1}/A_{i-1} \subseteq A_i/A_{i-1}$, and now the result follows from (i).

(iii) By proposition 3, A/H as an A-module has a fully reducible series of length $\leqq n$. By the correspondence between A-submodules and A/H-submodules of A/H, mentioned in the example (ii), this series is a fully reducible series of the A/H-module A/H which proves (iii) and the whole of corollary.

And the last result worth mentioning here is the

Proposition 4. *Let A be a Lie algebra and M be its irreducible module of finite length over k. Let also H be an ideal of A. Then any two chief factors of the H-module M are isomorphic.*

Proof. Assume that M contains a minimal H-submodule $N \neq \{0\}$ and consider a maximal nonzero H-submodule T, all of whose chief factors are isomorphic to N. We show that T is an A-submodule. For assume $xT \nsubseteq T$ for some $x \in A$. Then the mapping $t \mapsto xt + T$ gives rise to a homomorphism of the H-module T onto the H-module $xT + T/T$. Indeed, $ht \mapsto xht + T = hxt + [x, h]\, t + T = h(xt + T)$, since $H \trianglelefteq A$. By proposition 1 then all chief factors of $xT + T/T$ are isomorphic to those of T, that is to N. The reader is proposed now to prove that all chief factors of the H-module $xT + T$ are isomorphic to N. This, however, contradicts the maximality of T.

So far, T is an A-submodule and by our condition, $T = M$. The proof is complete.

CHAPTER 2

Free Lie algebras

2.1. The definition of a free Lie algebra

Given some set X, put $\Gamma_1(X) = X$ and define $\Gamma_n(X)$, $n > 1$, inductively as the set of all monomials of the form $(u)(v)$, $u \in \Gamma_s$, $v \in \Gamma_r$, $s + r = n$. (If, say, $s = 1$, write simply $u(v)$, or even uv, if also $r = 1$.) Having done this, put $\Gamma(X) = \bigcup_{n=1}^{\infty} \Gamma_n(X)$ and turn $\Gamma(X)$ into a groupoid subject to the operation $u \cdot v = (u)(v)$. An element $v \in \Gamma_s(X)$ is said to have *degree s* (denoted by $d(v) = s$). If X is fixed write Γ for $\Gamma(X)$ and Γ_s for $\Gamma_s(X)$.

Let further k be a commutative ring with unity, $F(X)$ be a k-algebra containing $\Gamma(X)$ as its free basis over k on the one hand and as a subgroupoid of its multiplicative groupoid on the other hand. In other words, any element from $F(X)$ has only one form $\sum_{u \in \Gamma} \alpha_u u$ and $\left(\sum_{u \in \Gamma} \alpha_u u\right)\left(\sum_{v \in \Gamma} \beta_v v\right) = \sum_{u,v \in \Gamma} \alpha_u \beta_v u \cdot v$. $F(X)$ is called a *free algebra* over k with set X of *free generators*. Clearly, any mapping of X into some k-algebra A uniquely extends to a homomorphism of $F(X)$ (or simply F) into A.

Denote by $F_n(X)$ (or simply F_n) the span of Γ_n in F, $n \geqq 1$. The elements from F_n are called *homogeneous* elements of degree n. Since $\Gamma_n \cap \Gamma_m = \emptyset$ if $m \neq n$, $F = \bigoplus_{n \geqq 1} F_n$. Obviously also $F_s F_r \subseteq F_{r+s}$ for all $r, s \geqq 1$. Any nonzero element $u \in F$ having the form $u = \sum_{i=1}^{n} u_i$, $u_n \neq 0$, n is called the degree of u, denoted by $d(u)$.

The concept of the degree in $\Gamma(X)$ (and $F(X)$) can be sharpened as follows. Let N_0 be the set of all nonnegative integers and $\mathsf{N}_0^{(X)}$ be the semigroup of all functions $\alpha : X \to \mathsf{N}_0$ with componentwise addition $(\alpha + \beta)(x) = \alpha(x) + \beta(x)$, $x \in X$, and finite support. Clearly, any nonzero element from $\mathsf{N}_0^{(X)}$ is a sum of a finite number of characteristic functions χ_x defined by $\chi_x(y) = 0$ if $x \neq y$ and $\chi_x(x) = 1$. Put $\Gamma_{\chi_x} = \{x\}$, for each $x \in X$, and if α is a nonzero function from $\mathsf{N}_0^{(X)}$ define inductively $\Gamma_\alpha(X)$ as a union $\Gamma_\alpha = \bigcup_{\beta+\gamma=\alpha} \Gamma_\beta \Gamma_\gamma$. Then, if $|\alpha| = \sum_{x \in X} \alpha(x)$, $\Gamma_n = \bigcup_{|\alpha|=n} \Gamma_\alpha$ and therefore $\Gamma = \bigcup_{\alpha \neq 0} \Gamma_\alpha$. An element from Γ_α is called a monomial of *multidegree* α. Clearly, each Γ_α is a finite subset of Γ.

Now denote by $F_\alpha = F_\alpha(X)$ a free k-submodule of $F(X)$ generated by Γ_α and call elements from F_α *multihomogeneous* of multidegree α. Given a characteristic function α of some finite subset $X_0 \subseteq X$, that is $\alpha(X_0) = 1$, $\alpha(X \setminus X_0) = 0$, an element from F_α is called *multilinear*.

For example, $(xy)\,y + (yx)\,y$ is a homogeneous but not a multihomogeneous element of $F(\{x, y\})$. On the contrary $2(xy)\,(zx) - (xx)\,(yz) - (zx)\,(yx)$ is a multihomogeneous element of multidegree $(2, 1, 1, 0)$ in $F(\{x, y, z, u\})$.

Note further that under any homomorphism of $F(X)$ into any associative algebra A all elements of the form $a(bc) - (ab)\,c$ $(a, b, c \in F)$, and thus the ideal B generated by these elements, map into zero. Moreover F/B itself is an associative algebra called *free associative algebra* over the set X. Similarly, the ideal C generated in F by all the elements of the form a^2 and $a(bc) + b(ca) + c(ab)$ $(a, b, c \in F)$, is contained in the kernel of any homomorphism of F into any Lie algebra. F/C is called a *free Lie algebra* over the set X. We shall identify in the sequel X with its images in F/B which is denoted by $A^+(X)$ and in F/C denoted by $L(X)$.

We show next that B and C are *multihomogeneous subspaces* of F in the sense that $B = \bigoplus_\alpha B \cap F_\alpha$ and $C = \bigoplus_\alpha C \cap F_\alpha$, that is each element from B (or C) is a sum of its multihomogeneous components which also lie in B (respectively in C).

We consider only the case of C, the case of B being left to the reader. Any element from C is a linear combination of elements having either the form

$$u_1 \dots u_m\big(a(bc) + b(ca) + c(ab)\big)\, v_1 \dots v_n\,, \tag{1}$$

or the form

$$u_1 \dots u_m(aa)\, v_1 \dots v_n\,, \tag{2}$$

where u_i, v_j be some monomials, $a, b, c \in F$, and the brackets being arranged in some way. Let $a = \sum_{i=1}^{s} a_{\alpha_i}$ be a multihomogeneous decomposition of a. Then (2) has the form

$$u_1 \dots u_m(a_{\alpha_i}a_{\alpha_j} + a_{\alpha_j}a_{\alpha_i})\, v_1 \dots v_n + \sum_{i=1}^{n} u_1 \dots u_m(a_{\alpha_i}a_{\alpha_i})\, v_1 \dots v_n\,. \tag{3}$$

Clearly, however $a_{\alpha_i}a_{\alpha_j} + a_{\alpha_j}a_{\alpha_i} = (a_{\alpha_i} + a_{\alpha_j})\,(a_{\alpha_i} + a_{\alpha_j}) - a_{\alpha_i}a_{\alpha_i} - a_{\alpha_j}a_{\alpha_j} \in C$ so that (3) is the sum of multihomogeneous elements which themselves lie in C. The case of (1) is dealt with similarly.

From the just proved property of B and C one easily sees that

$$F/B \cong \bigoplus_\alpha F_\alpha/F_\alpha \cap B \quad\text{and}\quad F/C \cong \bigoplus_\alpha F_\alpha/F_\alpha \cap C\,. \tag{4}$$

We put $F_\alpha/F_\alpha \cap B = A_\alpha$ and $F_\alpha/F_\alpha \cap C = L_\alpha$ and rewrite (4) in the form $A^+ = \bigoplus_\alpha A_\alpha$, $L = \bigoplus_\alpha L_\alpha$. Similarly if $F_n/F_n \cap B = A_n$, $F_n/F_n \cap C = L_n$, then $A^+ = \bigoplus_{n \geqq 1} A_n$ and $L = \bigoplus_{n \geqq 1} L_n$. By our convention, being now justified by $A_1 \cong F_1/F_1 \cap B \cong F_1 \cong L_1$, the submodules A_1 and L_1 are free k-submodules with free bases X. Further $A_n = \sum_{r+s=n} A_rA_s$, $L_n = \sum_{r+s=n} [L_r, L_s]$ (the operation in L will be denoted by bracketing), and the elements of A_n, L_n are called homogeneous of degree n. Finally $A_n = \bigoplus_{|\alpha|=n} A_\alpha$, $L_n = \bigoplus_{|\alpha|=n} L_\alpha$ and $A_\alpha = \sum_{\beta+\gamma=\alpha} A_\beta A_\gamma$, $L_\alpha = \sum_{\beta+\gamma=\alpha} [L_\beta, L_\gamma]$.

Just like $F(X)$, $A^+(X)$ and $L(X)$ possess *universal properties* respectively in the classes of all associative and Lie algebras over k. For example, given any map of X into some Lie algebra K over k, there exists a unique homomorphism of $L(X)$ into K extending this map.

In particular, the identity map of X extends to a homomorphism of $L(X)$ into $A^+(X)$ considered as a Lie algebra under bracketing. The injectivity of this homomorphism not being yet proved, we shall, however, identify $L(X)$ with its image under the homomorphism. Note that under this identification $L_\alpha \subseteq A_\alpha$ for all $\alpha \in \mathbb{N}_0^{(X)}$.

2.2. Bases of free Lie algebras

Our aim in this section is to prove that for any k, $L(X)$ is a free k-module and to point out explicitly some bases of $L(X)$.

The analogous question for $A^+(X)$ is being answered in much more simple way. Indeed, let $W(X)$ be the set of all nonempty words in the alphabeth X, $B(X)$ be the free k-module with the free basis $W(X)$ endowed with the multiplication whose restriction on $W(X)$ is juxtaposition in this latter semigroup. Clearly, $B(X)$ generated by X possesses the universal property in the class of all associative algebras over k, and hence the identity mapping of X extends to the isomorphism of $A^+(X)$ onto $B(X)$. Identifying $A^+(X)$ and $B(X)$ by this isomorphism, $A^+(X)$ can be regarded as a free k-module freely generated by all the monomials of the form

$$\{x_1 x_2 \dots x_s,\, s \geqq 1,\, x_i \in X,\, i = 1, 2, \dots, s\}\,.$$

To deal with $L(X)$, we introduce the concept of a "*basic set* of monomials" in $L(X)$. This is a totally ordered subset R of $\Gamma(X) \subseteq L(X)$ which is a union of its subsets $R_i = R \cap \Gamma_i$, $i = 1, 2, \dots$, such that (1) $R_1 = X$; (2) $w = [u, v] \in R_n$ if and only if $u \in R_m$, $v \in R_l$, $m + l = n$, and (i) $u < v$; (ii) if $v = [v_1, v_2]$ then $u \geqq v_1$; (3) $[u, v] > u$.

One easily sees that basic sets R exist (one may construct them using induction by n), the elements of R being called *basic monomials* of $L(X)$.

Lemma 1. *Let R be a basic set in L, $R_\alpha = R \cap \Gamma_\alpha$. Then L_α is the span of R_α, $\alpha \in \mathbb{N}_0^{(X)}$, $\alpha \neq 0$.*

Proof. We proceed by induction over $|\alpha|$, the case $|\alpha| = 1$ being trivial, and assume that w is an arbitrary monomial from $\Gamma_\alpha \subset L_\alpha$, $|\alpha| > 1$. Then $w = [u, v]$, $u \in \Gamma_\beta$, $v \in \Gamma_\gamma$, $\beta + \gamma = \alpha$. Since by the induction hypothesis u, v can be represented as linear combinations of elements from R_β, R_γ, assume without loss of generality that u and v are themselves basic and that, moreover, $u < v$. Include further into the induction hypothesis the assumption that any monomial of the form $[u, v]$, $u, v \in R$, $u < v$, is representable as a linear combination of basic monomials $w_1, w_2, \dots, w_s$ such that $w_i > u$, $i = 1, 2, \dots, s$. This obviously holds when $|\gamma| = 1$. Assume therefore $v = [v_1, v_2]$, $v_1 < v_2$, $v_1, v_2 \in R$; if $v_1 \leqq u$ then the proof is complete, so let $v_1 > u$. Then, by the Jacobi identity,

$$w = [u, [v_1, v_2]] = -[v_1, [u, v_2]] - [v_2, [u, v_1]]\,.$$

By the induction hypothesis $[u, v_2]$ and $[u, v_1]$ $(u < v_1 < v_2)$ are representable as linear combinations of basic monomials u' with $u' > u$, hence w is a linear combination of monomials of the form $[u', v']$, $u', v' \in R$, $u < u' < v'$, so that $[u', v'] \in \Gamma_\alpha \setminus \{[u, v]\}$. Similarly, for such a monomial if it is not basic one can find a representation as a linear combination of monomials $[u'', v'']$, $u'', v'' \in R$, $u' < u'' < v''$, so that $[u'', v''] \in \Gamma_\alpha \setminus \{[u, v], [u', v']\}$. But Γ_α is finite and therefore for some integer $s \geqq 1$, w is a linear combination of basic monomials $[u^{(s)}, v^{(s)}] \in R_\alpha$, $[u^{(s)}, v^{(s)}] > u^{(s)} \geqq u$ thus justifying our induction assumption. This completes the proof of the lemma.

Lemma 2. *Let B be a Lie subalgebra of some associative algebra C, the latter being generated by B as an associative algebra. Having assumed that B is a k-span of some totally ordered set E, C can be spanned by the set*

$$\{e_1 e_2 \dots e_n,\ e_1 \geqq e_2 \geqq \dots \geqq e_n,\ e_1, e_2, \dots, e_n \in E,\ n = 1, 2, \dots\}\,. \tag{1}$$

Proof. Clearly, C is a k-span of the set of all monomials of the form $w = f_1 f_2 \dots f_m$, $f_1, f_2, \dots, f_m \in E$. Denote by $i(w)$ the number of pairs (f_i, f_j) in w such that $f_i < f_j$ and $i < j$ at the same time. Using induction by m and, when m is fixed, by $i(w)$, we prove that w is a linear combination of the elements of the form (1) with $n \leqq m$; the base of the induction being supplied by the trivial cases $m = 1$ or $i(w) = 0$. Therefore assume $m > 1$ and $i(w) > 0$, that is there exists an i for which $f_i < f_{i+1}$ $(1 \leqq i \leqq m-1)$, and write $f_1 \dots f_i f_{i+1} \dots f_m = f_1 \dots f_{i+1} f_i \dots f_m + f_1 \dots [f_i, f_{i+1}] \dots f_m = w_1 + \sum_j \lambda_j f_1 \dots f_j \dots f_m$ where $[f_i, f_{i+1}] = \sum_j \lambda_j f_j$, $\lambda_j \in k$. Since $i(w_1) = i(w) - 1$ and since all the rest summands have degree $m - 1 < m$, the assertion of the lemma follows.

We use this lemma in the case $C = A^+(X)$ and $B = L(X)$, the image of a free Lie algebra in $A^+(X)$ generated by some set R of basic monomials as a k-module. Now by the preceding lemma, and also by lemma 1, for each $\alpha \in \mathsf{N}_0^{(X)}$, $\alpha \neq 0$, $A_\alpha(X)$ is spanned by the monomials of the form

$$w_1 w_2 \dots w_s,\quad w_i \in R_{\alpha_i},\quad \alpha_1 + \dots + \alpha_s = \alpha,\quad s \geqq 1,\quad w_1 \geqq \dots \geqq w_s\,. \tag{2}$$

On the other hand $A_\alpha(X)$, by the remark preceding lemma 1, is a free k-module freely spanned by

$$x_1 x_2 \dots x_n\,,\qquad n = |\alpha|\,,\qquad \chi_{x_1} + \dots + \chi_{x_n} = \alpha\,. \tag{3}$$

Our next aim is to show that the number of elements in (2) and (3) coincide. To do this we prove the last lemma in this section:

Lemma 3. *Let $\varphi\colon \Gamma(X) \to W(X)$ be the homomorphism of groupoids induced by the identity map of X, $W(X)$ being the semigroup of nonempty words in the alphabeth X. Then for each $w \in W(X)$, $\varphi^{-1}(w)$ contains exactly one monomial of the form*

$$\big(w_1(w_2(\dots(w_{s-1}w_s)\dots))\big)\,,\qquad w_i \in R,\quad w_1 \geqq w_2 \geqq \dots \geqq w_s,\quad s \geqq 1\,. \tag{4}$$

Proof. (i) Existence of (4) in $\varphi^{-1}(w)$ is proved by induction on the length $l(w)$ of w, the case $l(w) = 1$ being trivial. Assume thus $l(w) > 1$, that is $w = x_1 x_2 \dots x_s$, $x_i \in X$, $1 \leqq i \leqq n$, $n > 1$. Let $w' = x_1 x_2 \dots x_{n-1}$ and let (4) belong to $\varphi^{-1}(w')$. Assume then that $x_n \leqq w_s$; put $w_{s+1} = x_n$ and then clearly $w_1\big(w_2(\dots(w_{s-1}(w_s w_{s+1}))\dots)\big) \in \varphi^{-1}(w)$ and satisfies the inequalities $w_1 \geqq w_2 \geqq \dots \geqq w_s \geqq w_{s+1}$, so that the lemma is proved. More delicate is the case $x_n > w_s$. Let r be the least, if any, number such that $w_r < w_{r+1}\big(w_{r+2}(\dots(w_s x_n)\dots)\big)$ $(= [w_{r+1}, [w_{r+2}, [\dots[w_s, x_n]\dots]]])$. Then $w_r' = w_r(w_{r+1}$ $(\dots(w_s x_n)\dots))$ is a basic monomial with $w_{r-1} \geqq w_r'$ so that $w_1\big(w_2(\dots(w_{r-1}w_r')\dots)\big)$ has the form (4) and clearly $\in \varphi^{-1}(w)$.

(ii) Uniqueness of (4) in $\varphi^{-1}(w)$ is also obvious in the case $l(w) = 1$, so w may be assumed to have the form $w = x_1 x_2 \dots x_n$, $x_i \in X$, $1 \leqq i \leqq n$, $n > 1$. Let

$$\overline{w}_1\big(\overline{w}_2(\dots(\overline{w}_{t-1}\overline{w}_t)\dots)\big)\,,\qquad \overline{w}_j \in R\,,\qquad \overline{w}_1 \geqq \overline{w}_2 \geqq \dots \geqq \overline{w}_t\,,\qquad t \geqq 1\,, \tag{5}$$

be another element from $\varphi^{-1}(w)$ having the form (4).

Remark that any element from $\Gamma(X)$ is uniquely representable in the form

$$v_1(v_2(\dots(v_q x)\dots)), \qquad v_1, \dots, v_q \in \Gamma(X), \qquad x \in X. \tag{6}$$

If, moreover, (6) is an element of R, $v_1 \geqq v_2 \geqq \dots \geqq v_q < x$. Consider this decomposition for w_s and $\overline{w}_t$ from (4) and (5)

$$w_s = v_1(v_2(\dots(v_q x_n)\dots)), \qquad \overline{w}_t = \bar{v}_1(\bar{v}_2(\dots(\bar{v}_p x_n)\dots)).$$

Then the elements

$$w_1(w_2(\dots(w_{s-1}(v_1(v_2(\dots(v_{q-1}v_q)\dots))))\dots))$$

and

$$\overline{w}_1(\overline{w}_2(\dots(\overline{w}_{t-1}(\bar{v}_1(\bar{v}_2(\dots(\bar{v}_{p-1}\bar{v}_p)\dots))))\dots))$$

are easily seen to have the form (4) (use conditions (2) and (3) from the definition of the basic set R) and to belong to $\varphi^{-1}(w')$, $w' = x_1 x_2 \dots x_{n-1}$. Hence $w_1 = \overline{w}_1, \dots, v_q = \bar{v}_p$. Denote the sequence $w_1, \dots, v_q$ by $u_1, u_2, \dots, u_{s+q}$. Then by our definition q is the largest integer l for which

$$u_{s+q-l} < u_{s+q-l+1}(\dots(u_{s+q-1}(u_{s+q}x_n))\dots),$$

the same being true for p. Therefore $p = q$, $s = t$ and

$$w_1 = \overline{w}_1, \dots, w_s = v_1(v_2(\dots(v_q x_n)\dots)) = (\bar{v}_1(\bar{v}_2(\dots(\bar{v}_p x_n)\dots))) = \overline{w}_t.$$

This completes the proof of the lemma.

Clearly this lemma proves the coincidence of the number of generators of A_α having the form (2) and those ones having the form (3).

Remark further that if two systems $a_1, a_2, \dots, a_s$ and $b_1, b_2, \dots, b_s$ of elements of a free k-module M span this latter, the former being a free basis of M, then the latter one is as well free. Indeed, let A be a matrix from k_s such that $(a_1, a_2, \dots, a_s) A = (b_1, b_2, \dots, b_s)$ and B be such that $(b_1, b_2, \dots, b_s) B = (a_1, a_2, \dots, a_s)$. Then $AB = I_s$, the unity of a matrix ring k_s, whence $BA = I_s$. Given now $\lambda_1, \lambda_2, \dots, \lambda_s \in k$ with $\lambda_1 b_1 + \lambda_2 b_2 + \dots + \lambda_s b_s = 0$ one has $0 = (b_1, b_2, \dots, b_s)\begin{pmatrix}\lambda_1\\ \lambda_2\\ \vdots\\ \lambda_s\end{pmatrix} = (a_1, a_2, \dots, a_s) A \begin{pmatrix}\lambda_1\\ \lambda_2\\ \vdots\\ \lambda_s\end{pmatrix}$.
Since $(a_1, a_2, \dots, a_s)$ is free, $0 = A\begin{pmatrix}\lambda_1\\ \vdots\\ \lambda_s\end{pmatrix} = BA\begin{pmatrix}\lambda_1\\ \vdots\\ \lambda_s\end{pmatrix} = \begin{pmatrix}\lambda_1\\ \vdots\\ \lambda_s\end{pmatrix}$ which proves that $\{b_1, b_2, \dots, b_s\}$ is also k-free.

By this remark and using the direct decomposition $A^+ = \bigoplus_{\alpha \neq 0} A_\alpha$ of A as a k-module one sees now that the system of all monomials of the form

$$w_1 w_2 \dots w_s, \quad w_i \in R, \qquad w_1 \geqq w_2 \geqq \dots \geqq w_s, \quad s \geqq 1, \tag{7}$$

form a free k-basis of $A^+ = A^+(X)$. A part of this system consisting of monomials (7) with $s = 1$ generates $L = L(X)$ as a k-submodule of $A(X)$. This immediately implies that the natural mapping of $L(X)$ is injective and thus we have proved

Theorem 1 (Širšov). *Any set R of basic monomials of a free Lie algebra $L(X)$ over some commutative ring k forms a set of free generators of $L(X)$ as a free k-module. The canonical homomorphism of $L(X)$ into $A^+(X)$, the free associative algebra over k, induced by the identity map of X is injective.*

Concluding this section are two particular cases of the general scheme of basic monomial systems. First of them is one of the so-called '*Hall's bases*'. They may be obtained if the restriction on the ordering of $R(X)$ given by (3) is changed by a stronger one: given $w_1, w_2 \in R$ with $d(w_1) < d(w_2)$, $d(w_i)$ being the degree of w_i in X, one has $w_1 < w_2$.

To explain the second particular case due to Širšov (*Širšov's basis*) we need some more definitions. Let $S(X)$ be the semigroup of the words in the alphabeth X with lexicographical ordering. Namely, assuming X totally ordered, say that $x_s \dots x_2x_1 > x'_{s'} \dots x'_2x'_1$ if for no $t \geqq 1$ one has $x_1 = x'_1, \dots, x_{t-1} = x'_{t-1}, x_t < x'_t$. Thus, for example, if $x_1 < x_2 < x_3$, then $x_2x_3 > x_1x_2^2x_3 > x_2x_1x_3$. A word $a \in S(X)$ is called *basic* if $a = a_1a_2$ implies $a_2a_1 < a$. Let now $w \in \Gamma(X)$ be a nonassociative monomial and $\bar{w} = \varphi(w)$ (cf. lemma 3) be the corresponding word from $S(X)$. Call w *basic* if either $w \in X$ or the following two conditions hold:

(i) $w = [u, v]$ implies u, v are basic and $\bar{u} < \bar{v}$,

(ii) $v = [v_1, v_2]$ implies $\bar{u} \geqq \bar{v}_1$.

We denote by $\check{R}$ the set of all basic monomials arising in this way and prove that the relation $u < v \Leftrightarrow \bar{u} < \bar{v}$ is a total ordering of $\check{R}$ satisfying $u < [u, v]$, where $u, v, [u, v] \in \check{R}$. Then by theorem 1, $\check{R}$ is a basis of $L(X)$.

We prove first that if $w \in \check{R}$ then $\bar{w}$ is a basic word uniquely determining w. Indeed, one may proceed using induction over the degree of w in X. Suppose $\bar{w} = ax_1b$, x_1 being the least letter from X occurring in $\bar{w}$, b does not contain x_1. Clearly b is a nonempty word and if, moreover, $b = x_2b'$ for some $x_2 > x_1$, $[x_1, x_2]$ being a factor of w. Putting y for this $[x_1, x_2]$ into w and using the induction hypothesis (we preserve for y the order of x_1x_2 in S) one sees that $\bar{w}$ as a word in the new alphabeth is basic and that w is a single basic monomial whose associative word is $\bar{w}$. The same is true also for these words and monomials considered in the old alphabeth X, for the only thing remaining to mention is an evident inequality $ax_1b = ax_1x_2b' > x_2b'ax_1 = bax_1$. Practically the same argument shows that any basic word is an associative word of some basic monomial from $\check{R}$.

To finish the proof one clearly has to note that $\bar{u} \leqq \overline{uv}$ provided that $\bar{u} \leqq \bar{v}$ and $\bar{u}, \bar{v}, \overline{uv}$ are some basic (associative) words. Suppose, on the contrary, $\bar{u} > \overline{uv}$. This is possible only if $\bar{u} = \overline{zv}$ for some $\bar{z} \in S(X)$. Easily then $\bar{u} > \overline{uv}$ is equivalent to $\overline{zv} > \overline{uv}$ and $\bar{z} > \bar{u} = \overline{zv}$. From here either $\overline{vz} > \overline{zv}$ which is impossible since $\bar{u}$ is basic, or $\bar{u} = \overline{yz}$ and hence $\bar{u} = \overline{yz} > \overline{zy} > \overline{zv} = \bar{u}$, a contradiction, the latter inequality being true because $\bar{u} = \overline{zv} = \overline{yz} < \overline{vz}$ yields $\bar{y} > \bar{v}$ and because $\bar{y}$ and $\bar{v}$ have equal degrees. Now theorem 1 applies and $\check{R}$ is a basis of $L(X)$.

For example, basic monomials corresponding to x_2x_3, $x_1x_2^2x_3$ and $x_2x_1x_3$ are $[x_2, x_3]$, $[[[x_1, x_2], x_2], x_3]$ and $[x_2, [x_1, x_3]]$.

As a result one has

Proposition 1. *Let $\check{R} = \check{R}(X)$ be the set of all basic monomials obtained by arranging brackets in the basic words $\in S(X)$ as it is shown above. Then $\check{R}$ is a linear basis of a free k-module $L(X)$. Considering $L(X)$ as a subset of $A^+(X)$ one may represent uniquely any $w \in \check{R}$ in the form $w = \bar{w} + \sum_i \alpha_i a_i$, $\alpha_i \in k$, a_i are pairwise different words from $S(X)$ such that $a_i < \bar{w}$ for all i.*

Proof. The only thing to be checked up is the last statement which, however, may be obtained immediately proceeding by induction on the degree of w and taking into account the obvious fact that all a_i may be chosen of the same degree as w.

2.3. Subalgebras of free Lie algebras

Let k be a commutative ring with unity, X be some nonempty set, $L(X)$ be a free Lie algebra over k, $\Gamma(X)$ be a free groupoid identified with its image in $L(X)$. Let $S \subseteq \Gamma$ be a subset such that if $u \in \Gamma$, $v \in S$ then $[u, v] \in S$. Let further R be a basic set of monomials such that $u \in R \cap S$ and $v \in R \setminus S$ implies $u \geqq v$. An element $w \in S$ is called *S-reducible*, if $w = [u, v]$, $u, v \in S$, and *S-irreducible* otherwise.

Lemma 1. *Let Y be the set of S-irreducible elements from $R \cap S$. Then the submodule M spanned by $R \cap S$ is a free Lie algebra freely generated by Y.*

Proof. Consider $L(Y)$; it suffices to find a set R' of basic monomials in $\Gamma(Y)$ and an order-preserving imbedding of R' into R extending the identity map of Y. Denote by $R'_{(n)}$ the set $\bigcup_{i=1}^{n} R'_i$ and assume that $R'_{(n)}$ is already imbedded in R as an ordered set so that $R'_{(1)} = Y \subseteq L(X)$. Define $R'_{(n+1)}$ as the set of all products $[u, v], u, v \in R'_{(n)}$, such that the conditions 2,(i) and (ii) (from the preceding section) hold. Given $u, v \in R'_{(n)}$ consider $w = [u, v] \in R'_{(n+1)}$ as a corresponding element of $\Gamma(X)$. Then since the conditions defining basic sets hold for $R'_{(n)}$, w is easily seen to belong to R, for if v is S-reducible, this follows by the induction hypothesis, if, on the contrary, v is S-irreducible but $v = [v_1, v_2]$, $v_1 < v_2$ implies $v_1 \notin S$ so that $u > v_1$.

Thus $R'_{(n+1)} \subset R \cap S$. This induces an ordering in $R'_{(n+1)}$ possessing the property (3) from the definition of the basic set. Clearly also $R' \supseteq R \cap S$, since if w is S-irreducible, $w \in R'$, otherwise use induction by the degree of w, represented as a product of two elements from $R \cap S$.

Now the identity mapping of Y induces a homomorphism of $L(Y)$ into $L(X)$ under which the basic set R' of $L(Y)$ maps injectively onto the subset $R \cap S$ of the basic set R of $L(X)$. Thus on one hand, this homomorphism is injective, on the other hand the linear span of $R \cap S$ is a subalgebra isomorphic to $L(Y)$. The proof is complete.

Lemma 2. *Let $L(X)$ be a free Lie algebra over some ring k.*

(i) *Assume that $X = X' \cup X''$, $X' \cap X'' = \emptyset$; if $T = \mathrm{id}_L(X')$ is an ideal of L generated by X', then T is a free Lie algebra freely generated by the set of all elements either from X' or having the form*

$$[v_1, [v_2, [\dots [v_{s-1}, x_s] \dots]]], \tag{1}$$

$x_s \in X'$, $v_1 \geqq v_2 \geqq \dots \geqq v_{s-1}$, $v_1, v_2, \dots, v_{s-1} \in R''$, where R'' is a basic set of $L(X'')$ being in some way totally ordered, $s \geqq 2$.

(ii) *Assume that X is in some way totally ordered. Then $L(X)^2$ is either $\{0\}$ if $|X| \leqq 1$, or is a free Lie algebra freely generated by the monomials having the form*

$$[x_1, [x_2, [\dots [x_{s-1}, x_s] \dots]]], \tag{2}$$

$x_i \in X$ and $x_1 \geqq x_2 \geqq \dots \geqq x_{s-1} < x_s$, $s \geqq 2$.

Proof. In both cases we apply lemma 1.

(i) Denote by S the set of all monomials from $\Gamma(X)$ which lie in $\Gamma_\alpha, \alpha(X') \neq \{0\}$. Then clearly $u \in \Gamma$, $v \in S$ implies $[u, v] \in S$. Moreover, since the conditions '$u < [u, v]$' and '$u \in R \cap S$, $v \in R \setminus S$ implies $u \geqq v$' do not contradict each other a set R of basic monomials can be constructed satisfying to both of these conditions and thus lemma 1 applies. Obviously S-irreducible monomials from $R \cap S$ have the form (1).

(ii) In this case S is the set $\Gamma \setminus \Gamma_1$. The completion of the proof is left to the reader.

Let further G be a *graded* Lie algebra over k, that is $G = \bigoplus_{t=1}^{\infty} G_t$, $G_t, t = 1, 2, \ldots$, is a k-submodule of G, and $[G_t, G_s] \subseteq G_{s+t}$.

An element $x \in G_t$ is called a *homogeneous* element of *degree* t. Just like in section 1, the subalgebra H of G is called homogeneous if $H = \bigoplus_{t=1}^{\infty} H \cap G_t$. If G is a free Lie algebra then an example of a graduation of L (by the degrees of elements in the generating set X) was given in section 1. Remark, that in the next theorem L_t not necessarily means the set of elements of degree t with respect to the generating set X.

Theorem 2. *Let k be a commutative ring with* 1 *such that all projective k-modules are free*[1]) *and $L = L(X)$ be graded in some way free Lie algebra, X being homogeneous in this graduation. Let further H be a homogeneous subalgebra of L which is a direct summand of L as a k-module. Then H is a free Lie algebra with some set of homogeneous free generators.*

Proof. Let $L_t, t = 1, 2, \ldots$, be the homogeneous components of the above-mentioned graduation of L. Then $L = \bigoplus_{t=1}^{\infty} L_t$; if $H_t = L_t \cap H$, $H = \bigoplus_{t=1}^{\infty} H_t$; moreover, since for some free submodule K, $L = H \oplus K$ and $K \cong \bigoplus_{t=1}^{\infty} L_t/H_t$, all L_t/H_t are free, so that $L_t = H_t' \oplus L_t'$, H_t, L_t' being therefore free submodules of a free submodule L_t.

Note that the following reduction is admissible. Let $H^{(1)} = H_1 \bigoplus_{t=2}^{\infty} L_t$, $H^{(2)} = H_1^{(1)} \oplus H_2 \bigoplus_{t=3}^{\infty} L_t, \ldots, H^{(i+1)} = \bigoplus_{t=1}^{i} H_t^{(i)} \oplus H_{i+1} \bigoplus_{t=i+2}^{\infty} L_t, \ldots$. Assume that the subalgebra $H^{(i)}$ possesses a homogeneous system $B^{(i)}$ of free generators as a free Lie algebra such that $B_t^{(i)} = B_t^{(i-1)}$, $t = 1, 2, \ldots, i-1$. Then the system $B = \bigcup_{t=1}^{\infty} B_t^{(t)}$ is the desired system of free generators for H. For: any nontrivial linear combination in a basic set $R \subseteq \Gamma(B)$ is in fact a relation among the basic monomials in a finite subset $B' \subset B$. On the other hand B' is a subsystem of a system of free generators of some free subalgebra $H^{(i)}$ with suitable i. This contradiction proves our reduction.

So far, the only case to be dealt with is one of a subalgebra H such that $H_t = L_t$ for all $t \neq s$, s being a fixed integer, $s \geqq 1$. From the remark above, $L_s = H_s \oplus L_s'$. Write X_t for the set $X \cap L_t$. By the conditions of the theorem $X = \bigcup_{t=1}^{\infty} X_t$. Our goal is to find a system Y of homogeneous free generators of H such that $Y_t = X_t$ for all $t = 1, 2, \ldots, s-1$.

Given a (homogeneous) subalgebra G of L generated by $\bigcup_{t=1}^{s} X_t$, consider the k-submodule G_s. Then, since X is free, $L_s = kX_s \oplus G_s$. Clearly, $G_s \subseteq H_s$ and thus by Dedekind's law $H_s = (kX_s \cap H_s) \oplus G_s$. Now $kX_s/kX_s \cap H_s \cong kX_s + H_s/H_s = kX_s + G_s/H_s = L_s/H_s \cong L_s'$, since L_s' is free, $kX_s = (kX_s \cap H_s) \oplus L_s''$, where L_s'' is a free k-submodule of L which is isomorphic with L_s'. Let now $Z_s = Z_s' \cup Z_s''$ be another system of free generators of a free k-module kX_s such that $kZ_s' = kX_s \cap H_s$ and $kZ_s'' = L_s''$.

[1]) The reader not acquainted with the concept of a projective module may think of k as of a field or the ring $\mathbf{Z}$ of rational integers or more generally of a principal ideal ring.

Lemma 3. *Let $L(X)$ be a free Lie algebra over k, Z be a free basis of kX as a free k-module. Then $L(X) = L(Z)$.*

Proof. Clearly Z generates $L(X)$ and therefore to prove that Z is a free generating set of L it suffices to show that any linear combination of basic monomials in Z equal to zero is trivial. Let exist on the contrary such a nontrivial combination depending on a finite subset Z' of Z. Then for some finite $X' \subseteq X$, $kZ' \subseteq kX'$. As $kX = kZ = kZ' \oplus \oplus k(Z \setminus Z')$, by Dedekind's law,

$$kX' = kX' \cap kX = kZ' + k(Z \setminus Z') \cap kX' .$$

By our condition on k, the second summand of the right hand side is a free k-module freely generated by some Y. Now $k(Z' \cup Y) = kX'$. Denote $L(X')$ by L', and $Z' \cup Y$ by Y'. Then L' is generated by Y'. Moreover, any homogeneous component L'_n relatively to X is generated by basic monomials both in X' and Y' (since Y' is X'-homogeneous). Since the number of these monomials is finite and the same in both cases, the freeness of the former system of basic monomials yields that of the latter one (cf. lemma 2.3), and hence our lemma follows.

Using this lemma we may think without loss of generality that the set X of free generators of L is such that $X_s = X'_s \cup X''_s$, $kX_s = kX_s \cap H_s$, and $kX''_s = L''_s$. Let S be the set of all monomials either from $\bigcup_\alpha \Gamma_\alpha$, where $\alpha(x) \geqq 1$ for $x \in X \setminus X''_s$, or from $\bigoplus_{t>s} L_t$. Clearly S is the set satisfying the conditions of the preceding lemma 1, so that by this lemma for some set R of basic monomials from $\Gamma(X)$, the span of $S \cap R$, equal to H, is a free subalgebra freely generated by the set Y of S-irreducible monomials from $S \cap R$. One readily verifies that if $1 \leqq t \leqq s - 1$, then $Y_t = X_t$ and $Y_s = X'_s$.

Thus the proof of the theorem is complete.

Among the corollaries of this theorem the most popular are the following ones.

Theorem 4 (Širšov). *Let k be a field, L be the free Lie algebra over k with the set X of free generators, H be any subalgebra of L. Then H is a free Lie algebra over k.*

Proof. Given some graduation of L, say $L = \bigoplus_{t=1}^{\infty} L_t$, and a nonzero element $g \in L$ with homogeneous decomposition $g = g_1 + g_2 + \dots + g_s$, $g_s \neq 0$, call g_s a *major part* of g, denoted by $\bar{g}$. Denote also 0 by $\bar{0}$. Clearly then the set $\bar{H}$ of major parts of elements from H forms a homogeneous subalgebra of L which is a direct summand of L since k is a field. By theorem 3, $\bar{H} = L(\bar{T})$ for some subset $\bar{T} \subseteq \bar{H}$. We prove that $H = L(T)$. First note that if R is a basic set in $\Gamma(T)$ then $\bar{R}$ is a basic set in $\Gamma(\bar{T})$, since $\overline{w(t_1, t_2, \dots, t_r)} = w(\bar{t}_1, \bar{t}_2, \dots, \bar{t}_r)$. As $\bar{R}$ is linearly independent, the same, by the last equation, is true for R. We show finally, that $H = \text{alg}\,(T)$. Indeed, given $h \in H$, $\bar{h} \in \bar{H}_t$ for some $t \geqq 1$, and $\bar{h} = \sum_i w_i(\bar{t}_1, \dots, \bar{t}_r)$, $w_i \in L_t$. But then $h' = h - \sum_i w_i(t_1, \dots, t_r)$ has the major part $\bar{h}' \in L_{<t}$ and $h' \in H$; an evident induction thus completes the proof.

Theorem 5 (Witt). *Let L be a free Lie ring, that is an algebra over Z, H be its homogeneous subring.*

(i) *If L/H is a free abelian group, or*

(ii) *if $L = L(X)$, $|X| < \infty$, L/H is a torsion free abelian group,*

then H is a free Lie ring.

Proof. (i) immediately follows from theorem 3. Further, denoting by L_t the subspace of elements of L having degree t in X, one easily observes, that under the conditions of (ii) L_t is a finitely generated free abelian group whence L_t/H_t and therefore L/H are free abelian groups so that (i) applies.

Thus the proof of the theorem is complete.

Note also

Corollary. *Let L be a free Lie algebra over some field k, R be a proper nonzero ideal of L. Then R is an infinitely generated free Lie algebra.*

Proof. Since $R \neq L$, there exists an $x \in L \setminus R$. Consider $S = R \oplus kx$; this is a free subalgebra of L (by theorem 4) and $\dim S/R = 1$. Consider then a system Y of free generators of S, then $kY/kY \cap S$ has dimension 1. Choose a new basis Z of kY having the form $Z = \{x\} \cup Z', Z' \subset S$. Then $S = L(Z)$ by lemma 3. By lemma 2 (i), R is a free Lie algebra freely generated by all monomials of the form $z_{s,y} = (\mathrm{ad}(x))^s (y)$, $s \geqq 0$, $y \in Z'$. This completes our proof.

2.4. The universal enveloping algebra

Let G be a Lie algebra over some ring k with unity. An associative algebra B with unity 1 is called a *universal enveloping algebra* of G if there exists a homomorphism of Lie algebras $\varphi: G \to B$ (considered as a Lie algebra under bracketing) such that for any homomorphism of Lie algebras $\psi: G \to C$, C being some associative algebra with unity 1 over k, there exists a unique homomorphism of associative algebras $\chi: B \to C$ with $\chi(1) = 1$ and $\psi = \chi\varphi$.

Clearly the definition implies the uniqueness of the universal enveloping algebra (or, simply, the universal envelope), the existence being derived as follows. Denote by $A(X)$ the free associative algebra over k with unity 1, that is $A(X) = k\cdot 1 \oplus A^+(X)$. Then $A(X)$ is immediately seen to be the universal envelope of $L(X)$. Let G be a Lie algebra, then for some X, $G \cong L(X)/R$. Denote by S the least ideal of $A(X)$ containing R (L is considered as a Lie subalgebra of A generated by X). Then A/S is a universal envelope of G, for if ψ is a homomorphism of G into some associative algebra C, then ψ is induced by some homomorphism τ of L whose kernel contains R. Now τ induces some homomorphism π of associative algebras such that $\operatorname{Ker} \pi \supseteq S$. Denote by χ the homomorphism of A/S into B induced by π, then if φ is a (naturally arising) homomorphism $G \to A/S$, $\psi = \chi\varphi$ so that the definition of a universal envelope yields the desired.

The universal envelope of a Lie algebra G is denoted by $U(G)$.

Examples. (1) As we have seen already, $U(L(X)) \cong A(X)$.

(2) Given an abelian algebra G which is a free k-module, with free basis $x_1, x_2, \dots, x_n$. Then $U(G) \cong k[x_1, x_2, \dots, x_n]$.

Let G be a Lie algebra over k, $U(G)$ be its universal envelope, $\varphi: G \to U(G)$ be the homomorphism of Lie algebras from the definition. Call φ the *canonical map* of G into $U(G)$. It is convenient to denote by G the image of G in $U(G)$ under the canonical map, though we know nothing yet about the injectivity of this latter. Having this in mind, we prove the following

Proposition 1. *Let B be a Lie subalgebra of some associative algebra C generated by B as an associative algebra. Assume that E and F be two totally ordered free bases of B as a free k-module, k being a commutative ring over which all projective modules are free. Denote by S_E (or S_F) the set of all ordered monomials in E (respectively, F), that is, monomials having the form*

$$e_1e_2 \dots e_n\,, \qquad e_1 \geqq e_2 \geqq \dots \geqq e_n\,, \qquad n \geqq 1\,, \qquad e_i \in E$$

(respectively $f_1f_2 \dots f_m$, $f_1 \geqq f_2 \geqq \dots \geqq f_m$, $m \geqq 1$, $f_i \in F$). By the lemma 1.2 both S_E and S_F generate C as a k-module. Assume then that S_E is a free k-basis of the k-module C. Then S_F is as well a free basis of C.

Proof. Compare with the proof of lemma 3.3. Let D be some subset of B. Call an element $d_1d_2 \dots d_t$, $d_i \in D$, of C, a D-monomial of degree t, $t \geqq 1$. Denote by C_t the k-submodule of C spanned by all B-monomials of degree $\leqq t$, $t \geqq 1$, C_0 being a zero submodule of C. Clearly then $C = \bigcup_{t=0}^{\infty} C_t$. From the proof of lemma 1.2 one easily sees that C_t modulo C_{t-1} is spanned by E-monomials of degree exactly t. Moreover, having assumed S_E free, one immediately deduces that E-monomials of degree t form a free k-basis of C_t modulo C_{t-1}. Obviously, to prove our proposition it is enough to show that for each $t \geqq 1$, all F'-monomials of degree t form a free k-basis of C_t modulo C_{t-1}, F' being any finite subset of F. Let E' be a finite subset of E such that the k-submodule spanned by E' contains F', and F'' be a finite subset of F containing F' and such that the k-submodule spanned by F'' contains E'. Denote the k-spans of F', E', F'' respectively by U, V, W. Then $W = U \oplus T$, T being some free submodule of W. Using Dedekind's law one has $V = V \cap W = U \oplus (T \cap V)$. Since $T \cap V$ is a direct summand of the free k-module V, it is projective, hence, by our conditions, $T \cap V$ is free. Denote by E'' the union of F' and of a free basis of $T \cap V$. Then E' is a free k-basis of V and therefore $|E'| = |E''|$, moreover, the number of all ordered E'-monomials of degree t and the number of all ordered E''-monomials (E'' may be totally ordered, the order extending that of F') coincide. Clearly also the ordered E'- and E''-monomials of degree t generate the same free submodule modulo C_{t-1}; since the former system generates this submodule freely, the same is true for the latter one (see the remark afte lemma 1.3). Now as all ordered F'-monomials of degree t form a subset of a free system modulo C_{t-1}, they are also free modulo C_{t-1} and the proof is complete.

No we are in a position to prove the following fundamental

Theorem 6 (PBW). *Let k be a ring over which every projective module is free, G be a Lie algebra over k free as a k-module, $U(G)$ be its universal envelope, E be a totally ordered free k-basis of G. Then the canonical map of G into $U(G)$ is injective and $U(G)$ is a free k-module with a free k-basis formed by 1 and all ordered E-monomials*

$$e_1e_2 \dots e_s\,, \qquad e_1 \geqq e_2 \geqq \dots \geqq e_s\,, \qquad s \geqq 1\,. \tag{1}$$

Proof. Let first $G = L(X)$, a free Lie algebra over k, $E = R$, a basic system in $L(X)$. Then $A(X) = A_0 \oplus A^+(X)$, $A_0 \cong k1$, and lemmas 1, 2, 3 of the theorem 1 show that $A^+(X)$ is freely generated over k by the ordered R-monomials. Clearly this and the just proved proposition 1 provide the validity of the theorem for the case of a free Lie algebra.

Now given an arbitrary Lie algebra G, represent G as a quotient algebra $A = L(X)/R$ for suitable X. As we marked already, denoting then by S an ideal of $A(X)$ generated by R, one has $U(G) \cong A(X)/S$. Since G is a free k-module, $L = G' \oplus R$, $G' \cong G$, and by our conditions R is a free k-module. Let E' and E'' be two free k-bases of G', R, the image of E' under the natural epimorphism $\varepsilon: L \to G$ being equal to E. Introduce a total order in $E' \cup E''$ so that $E' \geqq E''$ and $\varepsilon|_{E'}$ is an isomorphism of ordered sets. Then by the first part of the proof, 1 and the ordered $(E' \cup E'')$-monomials

$$e_1'e_2' \dots e_s' \,, \quad e_1''e_2'' \dots e_l'' \,, \quad e_i' \in E' \,, \quad e_j'' \in E'' \,, \quad e_1' \geqq \dots \geqq e_s' \geqq e_1'' \geqq \dots \geqq e_l'' \quad (2)$$

form a free basis of $A(X)$ as a free k-module.

One immediately sees that the submodule of $A(X)$ spanned by the monomials (2) with $l > 0$ forms a two-sided ideal of A, which is contained and hence coincides with S. Then the basis of $A(X)/S$ is formed, by our convention on the identification of G with its image in $U(G)$, by the monomials (1). In particular, the elements of E are k-independent in $U(G)$, so that the canonical embedding is injective. This proves our theorem.

This theorem will be referred to as PBW-theorem in honour of H. Poincaré, G. Birkhoff and E. Witt whose contributions in it are mostly valuable.

Now denote by $U_n(G)$ or simply by U_n, $n \geqq 1$, the k-submodule of $U(G)$ spanned by all G-monomials of degree d, $1 \leqq d \leqq n$. Put also $U_0 = k \cdot 1$ and $U_{-1} = \{0\}$. Then obviously, $U_n U_m \subseteq U_{n+m}$, $n, m \geqq 0$. Consider then a k-module

$$gr(U(G)) = \bigoplus_{n=-1}^{\infty} U_{n+1}/U_n$$

and introduce the multiplication in $\mathrm{gr}(U(G))$ distributively and unambiguously by the rule $(g + U_{n-1})(h + U_{m-1}) = gh + U_{n+m-1}$, $g \in U_n$, $h \in U_m$. Using the proof of lemma 1.2, $\mathrm{gr}(U(G))$ is a commutative algebra over k with unity, graded by its submodules $\mathrm{gr}_n(G) = U_n/U_{n-1}$, $n = 0, 1, 2, \dots$. An immediate consequence from the PBW-theorem is the following

Corollary. *Under the conditions of the theorem, the identity mapping $E \to E$ extends to the isomorphism of $k[E]$, the polynomial ring over E, and* $\mathrm{gr}(U(G))$.

Remark. Let M be a G-module, G being a Lie algebra over k. Then M possesses a unique structure of a (left) $U(G)$-module compatible with that of a G-module. The converse is also true. To justify this, it suffices to consider the associative subalgebra A of $\mathrm{End}_k(M)$ generated by $g_M: x \mapsto gx$, $x \in M$, $g \in G$. Then $g \mapsto g_M$ is a homomorphism of G into A and hence by the universal property of $U(G)$ there exists a unique homomorphism of $U(G)$ into $\mathrm{End}_k(M)$ extending this mapping.

2.5. Free restricted Lie algebras

In this section we assume that the base ring k is a field of positive characteristic $p > 0$. Given a nonempty set X, a Lie p-algebra $L_p(X)$ is called a *free Lie p-algebra* (or a *free restricted* Lie algebra) with set X of free generators if $L_p(X)$ is generated by X and any mapping of X into some Lie p-algebra B extends to the unique homomorphism of $L_p(X)$ into B.

Proposition 1. *Let $A^+ = A^+(X)$ be the free associative algebra over k, $L = L(X)$ be its free Lie subalgebra generated by X. If R is a linear basis of L, then the set $R_1 = \{w^{p^t}, w \in R, t = 1, 2, \dots\}$ is linearly independent and spans a p-subalgebra $L_p(X)$, a free Lie p-algebra.*

Proof. The independence of R_1 follows from R_1 forming a part of a linear basis 4.(1) of the universal envelope A of L. Moreover, by proposition 1, 14.2 the span of R_1 is a p-subalgebra generated by $L(X)$ and thus by X.

Given further a map $\varphi: X \to B$, B being a Lie p-algebra, consider a homomorphism $\varphi_1: L(X) \to B$ extending φ. Since any element $a \in L_p(X)$ has the form

$$a = \alpha_1 w_1^{p^{t_1}} + \alpha_2 w_2^{p^{t_2}} + \dots + \alpha_s w_s^{p^{t_s}},$$

put

$$\bar{\varphi}(a) = \alpha_1 \varphi_1(w_1)^{p^{t_1}} + \alpha_2 \varphi_1(w_2)^{p^{t_2}} + \dots + \alpha_s \varphi_1(w_s)^{t_s}.$$

Using identities from section 1, 14, one immediately verifies that $\bar{\varphi}$ is a homomorphism, thus proving our proposition. The elements of R_1 are called *p-basic monomials* in X.

Free Lie p-algebras enjoy many properties of free Lie algebras and free groups at the same time. This can be maintained by the following two theorems.

Theorem 6 (Witt). *Let k be a field of positive characteristic $p > 0$, $L_p(X)$ be a free Lie p-algebra over k, H be some p-subalgebra of $L_p(X)$. Then H is a free Lie p-algebra.*

Proof. Let f be any element of $A^+(X) = \bigoplus_{n=1}^{\infty} A_n(X)$ having the decomposition $f = f_r + f_{r+1} \dots + f_s$, $r \leqq s$, $f_r, f_s \neq 0$, with respect to this sum. The summands f_r and f_s are termed as the *junior* and *major parts* of f, denoted by $\tilde{f}$ and $\bar{f}$ respectively. A set T of elements of $L_p(X)$ is called a *p-reduced set* if no major part $\bar{f}$ of an element f from T lies in the p-subalgebra $\mathrm{alg}_p(\{\bar{g}, g \in T \setminus \{f\}\})$ generated by the major parts of all the rest elements from T.

Lemma 1. *Any p-subalgebra B of a free Lie p-algebra $L_p(X)$ possesses a p-reduced set T of generators.*

Proof. Set $T_0 = \emptyset$. Assuming $T_0, T_1, \dots, T_i$ to be constructed, consider a p-subalgebra E_i generated by the set $\bigcup_{j=0}^{i} T_j$ and its subspace S_i spanned by the elements having degree $\leqq i + 1$. Denote then by T_{i+1} the maximal set of elements of B having degree $\leqq i + 1$ and being linearly independent modulo S_i. Evidently B can be generated by $T = \bigcup_{i=0}^{\infty} T_i$. Moreover, if $f \in T_{i+1}$, then $\bar{f}$ is a homogeneous element of the degree $i + 1$; if $\bar{f}$ were an element of the subalgebra C generated by the major parts of the rest elements from T then, since these parts are homogeneous, C should be generated by the major parts of the elements from $\bigcup_{s=0}^{i+1} T_s$. This, however, easily gives a contradiction with our choice of T_{i+1}, and the lemma is proved.

Let now $S = \{s_\alpha, \alpha \in I\}$ be a set of elements of a free Lie p-algebra $L_p(X)$. An *elementary transformation* of S is a map $\omega: S \to L_p(X)$ given by $\omega(s_\alpha) = s_\alpha$ for all $\alpha \in I$ but a fixed one, β, and $\omega(s_\beta) = \lambda s_\beta + w(s_{\alpha_1}, \dots, s_{\alpha_t})$, with $0 \neq \lambda \in k, \alpha_1, \dots, \alpha_t \neq \beta$ and $w(x_{\alpha_1}, \dots, x_{\alpha_t}) \in L_p(X)$. A set Z of elements of $L_p(X)$ will be called a *p-independent* set if Z is a set of free generators of a p-algebra $\mathrm{alg}_p(Z)$, generated by Z.

Lemma 2. *Given a free Lie p-algebra $L_p(X)$ and an elementary transformation ω of X, $\omega(X)$ is a p-independent set of generators of $L_p(X)$.*

Proof. Clearly $\omega(X)$ generates $L_p(X)$. Moreover, a nontrivial relation among the basic monomials in $\omega(X)$ implies a nontrivial relation among basic monomials in $\widetilde{\omega(X)}$, the set of junior parts of elements from $\omega(X)$. The remaining part of the proof is dealt with using lemma 3.3 and proposition 4.1.

Corollary 1. *Elementary transformations induce automorphisms of $L_p(X)$.*

Lemma 3. *Let $X = \{x_1, x_2, \dots, x_n\}$ be some p-independent set of a Lie p-algebra L. Then the set $X_1 = \{x_1^p, (\mathrm{ad}(x_1))^l (x_\alpha) \mid 1 < \alpha \leqq n, \ 0 \leqq l < p\}$ is p-independent.*

Proof. The method of the proof is the same as in the case of lemma 3.1. Any basic monomial in X_1 is either x_1^p or a basic monomial in the system 3.(1), if $X'' = \{x_1\}, x_1^p$ being the least element of X_1. Now, obviously, the union of these two systems is linearly independent, which proves the lemma.

Lemma 4. *Any p-reduced subset of a free Lie p-algebra $L_p(X)$ is p-independent.*

Proof. Suppose not. Any nontrivial relation holding in M implies a nontrivial relation in $\overline{M}$. Since $\overline{M}$, the set of the major parts of the elements of M, is also p-reduced, one may assume without loss of generality, that M is a homogeneous set. Moreover, evidently X and M may be supposed to have finite cardinalities.

Since the elements $f_1, f_2, \dots, f_l$ of degree 1 from M are linearly independent, there exists a nondegenerate linear map $\varphi\colon kX \to L_p(X)$ such that $\varphi(x_1) = f_1, \dots, \varphi(x_l) = f_l$. Having been represented as a product of elementary transformations, this map induces by the corollary of lemma 2, an automorphism φ of $L_p(X)$ which transforms the elements of X in the same way, and we identify X with $\varphi(X)$. Clearly, M is still p-reduced with respect to this new set of free generators. The cases $M \subseteq X$ or $\{x_1, x_2, \dots, x_l\} = X$ being trivial, consider the element $x_{l+1} \notin M$. Then M is a subset of the ideal I of $L_p(X)$ generated by $(X \setminus \{x_{l+1}\}) \cup \{x_{l+1}^p\}$, which is a free Lie p-algebra with a set of free generators Y given by lemma 3. Clearly, M is a p-reduced set in $L(Y)$ having total degree in Y strictly less than that in X. The evident induction now completes the proof of the lemma and the whole theorem 6.

And, finally, a theorem showing how near is the theory of free Lie p-algebras to that of free groups.

Theorem 7 (Kukin). *Let B be a p-subalgebra of a free Lie algebra $L_p(X)$ of rank n, that is, $|X| = n$. If $\dim (L/B) = j < \infty$ then B is a free Lie p-algebra of rank $N(n, j) = p^j(n - 1) + 1$.*

We start with

Lemma 5. *Let B be generated by a p-reduced set M. Then $\overline{B}$ generated by a p-reduced set $\overline{M}$ of major parts of elements from M has the same codimension in $L_p(X)$, that is* $\dim(L/B) = \dim(L/\overline{B})$.

Proof. Let $Q_0 = \emptyset$; $Q_0, Q_1, \dots, Q_{t-1}$ being already defined, denote by Q_t a maximal set of p-basic monomials having degree t and being linearly independent modulo the subspace of L_p spanned by B and $\bigcup_{i=0}^{t-1} Q_i$. Then $Q = \bigcup_{i=0}^{\infty} Q_i$ spans a linear subspace T of L_p

complementing B, that is $L_p = B \oplus T$. We show that also $L_p = \overline{B} \oplus T$. Assume first $\sum_{i=1}^{q} \lambda_i w_i = \psi(\overline{m}_1, \dots, \overline{m}_r)$, where all w_i may be supposed to have the same degree t, $w_i \in Q_i, m_j \in M$. Then $\sum_{i=1}^{q} \lambda_i w_i = \psi(m_1, \dots, m_r) + \varphi$, $\deg(\varphi) < t$; by our choice, $\varphi = \pi(m_1, \dots, m_r) + \sum_{j=1}^{s} \mu_j w_j$, $\deg(w_j) < t$ (to find such a decomposition, represent any basic summand of φ in such a form, using the definition of Q). Clearly this contradicts the choice of Q, so that our assumption is false and $\overline{B}$ and T are linearly independent. Similarly, one may show that $L_p = \overline{B} + T$. This proves our lemma.

Proof of the theorem. By the just proved lemma, M may be assumed to be homogeneous, moreover just like in the proof of lemma 4, one may assume that the elements of degree 1 of M form a part of X. Then if $x_1 \in X \setminus M$, B is a subalgebra of a free Lie algebra I generated by a p-reduced set $Y = \{x_1^p, \ (\mathrm{ad}(x_1))^l\ (x_\alpha) \mid 1 < \alpha \leqq n;\ 0 \leqq l < p\}$ (see lemma 3). Since $\dim\big(L(Y)/B\big) = j - 1$, by the induction hypothesis, B is a free p-algebra of rank $N\big((n-1)\,p + 1, j - 1\big) = p^{j-1}\big(p(n-1)\big) + 1 = p^j(n-1) + 1$.

This proves the theorem.

Corollary 2. *Let B be a p-subalgebra of a finitely generated Lie p-algebra A such that* $\dim(A/B) < \infty$. *Then B is finitely generated.*

2.6. Free products

Let $(G_\alpha)_{\alpha \in I}$ be a family of Lie algebras over some commutative ring k possessing free bases $B_\alpha = \{e_{\alpha\beta} \mid \beta \in I_\alpha\}$ over k. Then the multiplication in each G_α is given by the formulae

$$[e_{\alpha\beta}, e_{\alpha\gamma}] = \sum_\tau p_{\alpha\beta\gamma}^\tau e_{\alpha\tau}\,, \qquad \beta, \gamma \in I_\alpha\,, \tag{1}$$

where $p_{\alpha\beta\gamma}^\tau \in k$ are the "structure constants" and $\tau \in I_\alpha$. Put $X_\alpha = \{x_{\alpha\beta} \mid \beta \in I_\alpha\}$ and consider the set $X = \bigcup_\alpha X_\alpha$. Let further $L = L(X)$ be the free Lie algebra over k with the set X of free generators and J be its ideal generated by the set $D = \{d_{\alpha\beta\gamma} \mid \alpha \in I, \beta, \gamma \in I_\alpha\}$, where

$$d_{\alpha\beta\gamma} = [x_{\alpha\beta}, x_{\alpha\gamma}] - \sum_\tau p_{\alpha\beta\gamma}^\tau\, x_{\alpha\tau}\,. \tag{2}$$

Then L/J is denoted by $\Pi^*_\alpha\, G_\alpha$ and called a *free product* of algebras G_α, $\alpha \in I$.

Proposition 1. *Let ε be the natural homomorphism $L \to P = \Pi^*_\alpha\, G_\alpha$. Then for each $\alpha \in I$ a subalgebra $\overline{G}_\alpha$ of P generated by $\varepsilon(X_\alpha)$ is isomorphic with G_α. If for each $\alpha \in I$, φ_α: $\overline{G}_\alpha \to G$ is a homomorphism of $\overline{G}_\alpha$ into some Lie algebra G over k, then there exists a unique homomorphism $\varphi\colon P \to G$ such that $\varphi|_{\overline{G}_\alpha} = \varphi_\alpha$.*

We prove first the second statement. Let $\psi\colon X \to G$ be a mapping defined by $\psi(x_\alpha) = \varphi_\alpha\big(\varepsilon(x_\alpha)\big)$, $\alpha \in I$. Then ψ extends to a homomorphism $\chi\colon L \to G$ whose kernel evidently contains each $d_{\alpha\beta\gamma}$ and consequently $\operatorname{Ker}\chi \supseteq J$. Then clearly there exists a homomorphism $\varphi\colon P = L/J \to G$ with the desired properties. Its uniqueness follows since P is generated by its subalgebras $\overline{G}_\alpha$, $\alpha \in I$.

The first part is now easy if one considers a homomorphism φ of a free product $\prod\limits_\alpha^* G_\alpha$ onto $Q = \prod\limits_\alpha G_\alpha$ induced by the mapping $\chi: L \to Q$ such that $\chi(x_{\alpha\beta}) = e_{\alpha\beta}$, $\alpha \in I$, $\beta \in I_\alpha$.

Thus the proof is complete.

Now we calculate the universal envelope $U(P)$, $P = \prod\limits_\alpha^* G_\alpha$ through the universal envelopes $U_\alpha = U(G_\alpha)$, $\alpha \in I$. An associative algebra $A = \prod\limits_\alpha^* A_\alpha$ is called a *free product* of associative algebras A_α, $\alpha \in I$, if $A \supseteq A_\alpha$ and for each $\mu_\alpha: A \to B$, B being some associative algebra, there exists a unique homomorphism $\mu: A \to B$ such that $\mu|_{A_\alpha} = \mu_\alpha$, $\alpha \in I$. If T_α is a basis of A_α over k, then the basis of A is formed by the monomials of the form

$$t_1 t_2 \dots t_q\,, \qquad t_i \in T_{\alpha_i}\,, \qquad \alpha_i \neq \alpha_{i+1}\,, \qquad q > 0\,. \tag{3}$$

The trivial proof of this fact is left to the reader.

This also immediately implies that $U^+(P) = \prod^* U^+(G_\alpha)$ (use proposition 1), where U^+ is the set of elements of U without constant term. By the PBW-theorem 4.6 a basis of U may be found then as follows.

Assume that I, I_α, $\alpha \in I$, are totally ordered and order $B = \bigcup\limits_\alpha B_\alpha$ by the rule $e_{\alpha\beta} > e_{\alpha'\beta'}$ if either $\alpha > \alpha'$, or $\alpha = \alpha'$ and $\beta > \beta'$. Then a basis of $U(P)$ has the form

$$e_{\alpha_1 i} \dots e_{\alpha_1 j} e_{\alpha_2 s} \dots e_{\alpha_2 r} \dots e_{\alpha_t n} \dots e_{\alpha_t m}\,, \qquad t \geqq 0 \tag{4}$$

and $\alpha_i \in I$, $i \geqq \dots \geqq j$, $s \geqq \dots \geqq r$, $\dots$, $n \geqq \dots \geqq m$, $\alpha_i \neq \alpha_{i+1}$, $i = 1, \dots, t-1$.

Using this basis of the universal envelope $U(P)$ and a basis $\check{R}$ of a free Lie algebra due to Širšov (see proposition 2.1) one may construct a basis of P as a free k-module in the way this is done below. Let $\check{R}(E)$ be a set of (Širšov) basic monomials in E. Call $w \in \check{R}$ a *special* basic monomial if $\bar{w}$ does not contain a subword of the form $e_{\alpha i} e_{\alpha j}$ with $i < j$.

Theorem 8 (Širšov). *The set R_s of all special basic monomials forms a basis of a free k-module $P = \prod^* G_\alpha$.*

Proof. Let $w \in R_s$; consider w as an element of $U = U(P)$ and represent it as the unique linear combination of the words having the form (4). Denote by $\hat{w}$ its lexicographically greatest summand of the same degree as that of w. By proposition 1, the PBW-theorem 4.6 and the definition of the special basic monomials, $\hat{w} = \bar{w}$, thus proving that the system R_s is linearly independent.

To complete the proof it suffices to express any $w \in \check{R}$ through R_s. To this end, consider $w \in \check{R}$ such that $\bar{w}$ has a subword $e_{\alpha i} e_{\alpha j}$ with $i < j$. Obviously then w possesses a factor of the form $[[w_t, \dots, w_2, w_1, e_{\alpha i}], e_{\alpha j}]$, where all w_i are basic monomials and $w_t \geqq \dots \geqq w_2 \geqq w_1 < e_{\alpha i}$. Let w' be a basic monomial obtained from w by replacing this factor by a monomial $[w_t, \dots, w_2, w_1, e_{\alpha i}, e_{\alpha j}]$. Then using proposition 2.1, w' is a linear combination of basic monomials of the same degree as w of the form $w' = w + \sum\limits_i \alpha_i u_i$, where $\bar{u}_i < \bar{w}$. By the induction hypothesis each u_i is a linear combination of special basic monomials and w' is a linear combination of monomials having strictly lower degree, for $[e_{\alpha i}, e_{\alpha j}] = \sum\limits_l p_{\alpha ij}^l e_{\alpha l}$.

This completes the proof of the theorem.

Many important subalgebras of free products of Lie algebras are known to be free. However proofs of such results are rather formidable because of the plenty of technical details. Therefore we limit ourselves to only one such a result (we shall need it in Chapter 3).

Proposition 2. *Let C be a free Lie algebra, B be any Lie algebra over the same field k, $P = C * B$ be their free product. Then the ideal $D = \mathrm{id}_P(C)$ is a free Lie algebra.*

Proof. Let $\{c_i, i \in I\}$ be a set of free generators of C and $\{e_\alpha, \alpha \in M\}$ be a basis of B over k, M being in some way totally ordered. We show that a set of free generators of D may be taken in the form

$$[e_{\alpha_1}, e_{\alpha_2}, \ldots, e_{\alpha_n}, c_i]\,, \qquad \alpha_1 \geqq \alpha_2 \geqq \ldots \geqq \alpha_n\,, \qquad n \geqq 0\,, \qquad i \in I\,. \tag{5}$$

To do this consider a free Lie algebra R freely generated by a set $\{(m, c_i), i \in I, m$ being any basic monomial of $U(B)\}$. Now we may construct a homomorphism $b \mapsto d_b$ of B into $\mathrm{Der}(R)$ by putting $d_b((m, c_i)) = \sum_{j=1}^{t} \alpha_j(m_j, c_i)$ if $bm = \sum_{j=1}^{t} \alpha_j m_j$ in $U(B)$. Each d_b uniquely extends to a derivation of R and, clearly, the map $b \mapsto d_b$ defined in this way is a homomorphism of B into $\mathrm{Der}(R)$ since this is true on the free generators of R. Form then a semidirect product $T = B \curlywedge R$ relatively to this homomorphism. By the definition of a free product there exists a homomorphism $\chi\colon P \to T$ induced by the identity mapping of B and by the mapping $c_i \mapsto (1, c_i)$ of C into R. Clearly, under this homomorphism the images of the elements having the form (5) are free generators of the subalgebra R, therefore the same is true for the subalgebra of D generated by (5). This subalgebra however coincides with D by lemma 3.2(i) and the proof of the proposition is complete.

CHAPTER 3

Varieties of Lie algebras

3.1. Identical relations in Lie algebras

Let k be a ring, X be a set, $L = L(X)$ be a free Lie algebra over k with a set X of free generators, A be some Lie algebra over k. Given a nonzero element $v \in L$ and a subset $\{x_1, x_2, \dots, x_n\} \subseteq X$ such that $v \in L(\{x_1, \dots, x_n\}) \subseteq L(X)$, one says that the *identical relation*

$$v \equiv 0 \quad \text{or} \quad v(x_1, x_2, \dots, x_n) \equiv 0 \tag{1}$$

holds in A (or that A *satisfies* (1)) if under any homomorphism $\varphi: L \to A$, $\varphi(v) = v(\varphi(x_1), \dots, \varphi(x_n)) = 0$. In other words, if $g_1, g_2, \dots, g_n$ are arbitrary elements of A, $v(g_1, g_2, \dots, g_n)$ is the *value* of v in $g_1, g_2, \dots, g_n$, that is the image $\varphi(v)$ of v under a homomorphism of L, induced by $x_i \mapsto v_i$, $i = 1, 2, \dots, n$, then (1) holds in A if all the values of $v(x_1, x_2, \dots, x_n)$ in A equal zero.

Let further

$$w \equiv 0 \quad \text{or} \quad w(y_1, y_2, \dots, y_m) \equiv 0 \tag{2}$$

be another identical relation, $w = w(y_1, \dots, y_m) \in L(Y)$ for some set Y. Say that relation (1) *implies* relation (2) if as soon as (1) holds in some Lie algebra A, so does (2). If also (2) implies (1), one says that (1) is *equivalent* to (2).

Proposition 1. *Let X be a countable set, Y be any set. Given any identical relation having form* (2), *there exists an equivalent identical relation having form* (1).

Proof. Let $x_1, x_2, \dots, x_m$ be elements from X, $w(x_1, x_2, \dots, x_m)$ be the value of $w(y_1, y_2, \dots, y_m)$ in $x_1, x_2, \dots, x_m$. Then the proposition immediately follows from the sequence $L(Y) \xrightarrow{\pi} L(\{y_1, y_2, \dots, y_m\}) \xrightarrow{\varepsilon} L(\{x_1, x_2, \dots, x_m\}) \xrightarrow{\iota} L(X) \xrightarrow{\varphi} A$ of homomorphisms of Lie algebras, induced in the natural way by the corresponding mappings of free generators, A being an arbitrary Lie algebra. Namely, π is an extension of some projection of Y onto $y_1, y_2, \dots, y_m$, ε is an extension of $y_i \mapsto x_i$, ι is an extension of $x_i \mapsto x_i$. The rest of the proof is left to the reader.

Thus, due to proposition 1, all the left hand sides of the identical relations may be considered as elements of some fixed Lie algebra $L(X)$, X being some fixed countable set.

Examples. (i) Any abelian Lie algebra satisfies an identical relation of the form

$$[x, y] \equiv 0 . \tag{3}$$

(ii) Any nilpotent Lie algebra of nilpotency class $d \leq c$ satisfies an identical relation

$$[x_{c+1}, x_c, \dots, x_2, x_1] \equiv 0 . \tag{4}$$

(iii) Set $\delta_0(x) = x$ and define inductively $\delta_{l+1}(x_1, x_2, \dots, x_{2^{l+1}})$ by the rule

$$\delta_{l+1}(x_1, x_2, \dots, x_{2^{l+1}}) = [\delta_l(x_1, x_2, \dots, x_{2^l}), \delta_l(x_{2^l+1}, x_{2^l+2}, \dots, x_{2^{l+1}})].$$

Then any solvable Lie algebra of solvable length $t \leq l$ satisfies an identical relation

$$\delta_l(x_1, x_2, \dots, x_{2^l}) \equiv 0. \tag{5}$$

(iv) Let A be some m-generator as a k-module Lie algebra over some ring k, S_{n+1} be the symmetrical group of degree $n + 1$. Then for any $n \geq m$ the following identical relation holds in A:

$$\sum_{\sigma \in S_{n+1}} |\sigma| \, [x_{\sigma(n+1)}, \dots, x_{\sigma(1)}, x_0] \equiv 0, \tag{6}$$

where $|\sigma| = 1$ if σ is even and $|\sigma| = -1$ if σ is odd.

All routine calculations are left to the reader.

Note that the classes of Lie algebras in the cases (i), (ii), and (iii) are completely determined by the corresponding identities (3), (4), and (5), while in the case (iv) this is not true. One distinguishes classes of the first kind and calls them varieties of Lie algebras. More explicitly, a class $\mathfrak{B}$ of Lie algebras over some commutative ring is called a *variety* if there exists some fixed set of identical relations V such that a Lie algebra A over k belongs to $\mathfrak{B}$ if and only if A satisfies all identities from V.

Thus one may speak on the variety $\mathfrak{A}$ of all abelian Lie algebras over k, on the variety $\mathfrak{N}_c$ of all nilpotent Lie algebra of class $\leq c$, on the variety $\mathfrak{S}_l$ of all solvable Lie algebras having solvable length $\leq l$, and so on.

An abstract characterization of varieties is given by the following theorem, which however is valid for far broader classes of algebraic systems.

Theorem 1 (Birkhoff). *A nonempty class $\mathfrak{B}$ of Lie algebras over some ring k is a variety if and only if $\mathfrak{B}$ is closed under taking of subalgebras, homomorphic images and cartesian products of its algebras.*

Proof. The 'only if' part being evident, it is enough to show that, given a closed class $\mathfrak{B}$ of Lie algebras over k and a Lie algebra A over k, satisfying all the identities holding in $\mathfrak{B}$, then $A \in \mathfrak{B}$.

To do this, represent, first, A as a quotient algebra of some free Lie algebra $L(Y)$ for some set Y. Separate, second, from $\mathfrak{B}$ any subclass $\mathfrak{W}$ which determines the same identities as $\mathfrak{B}$ and denote by $\mathfrak{U}$ the *set* of all nonisomorphic images of $L(Y)$ in $\mathfrak{W}$. Let then Φ be the set of all homomorphisms of $L(Y)$ in $\mathfrak{U}$ and

$$P = C \prod_{\varphi \in \Phi} \varphi(L).$$

Denote by f_y, $y \in Y$, an element from P such that $f_y(\varphi) = \varphi(y)$ and consider a subalgebra $B = \text{alg}(\{f_y, y \in Y\}) \in \mathfrak{B}$, generated by f_y, $y \in Y$, which is the homomorphic image of $L(Y)$ under the homomorphism χ, extending the mapping $y \mapsto f_y$, $y \in Y$.

Given now an epimorphism $\psi \colon L(Y) \to A$, we show that $\text{Ker}\,\psi \supseteq \text{Ker}\,\chi$, for then, by the theorem 1,3.1 (iii), A is a quotient algebra of B, that is $A \in \mathfrak{B}$. However, if $w \in \text{Ker}\,\chi$, then $\varphi(x) = 0$ for all $\varphi \in \Phi$ and therefore for all homomorphisms of $L(Y)$ into $\mathfrak{W}$. Thus $w \equiv 0$ is an identity in $\mathfrak{B}$ so that $\psi(w) = 0$, since A satisfies all such identities. This completes the proof of the theorem.

From the proof of this theorem one easily deduces the existence in each variety $\mathfrak{B}$ for each cardinal number $\mathfrak{m}$ of an $\mathfrak{m}$-generator algebra B such that any $\mathfrak{m}$-generator algebra A from $\mathfrak{V}$ is a homomorphic image of B. Moreover this algebra possesses a set Z of generators such that any map of Z into some Lie algebra $A \in \mathfrak{V}$ extends (uniquely) to a homomorphism of B into A. We shall denote this algebra B by $L(Z, \mathfrak{V})$ and call it a $\mathfrak{V}$-*free* (or *relatively free*) algebra with the set Z of free generators. Obviously $\operatorname{Card}(Y) = \operatorname{Card}(Z)$ implies $L(Y, \mathfrak{V}) \cong L(Z, \mathfrak{V})$ and justifies the notation $L(Y, \mathfrak{V}) = L_{\operatorname{Card}(Y)}(\mathfrak{V})$. Now we can deduce an important

Corollary. *Let A be a finite Lie algebra over k, $\mathfrak{V} = \operatorname{var}(A)$ be the variety of Lie algebras generated by A, that is the least variety of Lie algebras over k containing A. Then $\mathfrak{V}$ is locally finite, that is any finitely generated Lie algebra B from $\mathfrak{V}$ is finite.*

Proof. Since any n-generator algebra from $\mathfrak{V}$ is isomorphic to a quotient algebra of $L_n(\mathfrak{V})$ it suffices to prove that $B = L_n(\mathfrak{V})$ is finite. Take in the proof of the just proved theorem $\mathfrak{W} = \{A\}$, $Y = \{y_1, y_2, \dots, y_n\}$; then Φ is finite so that B is a subalgebra of a finite direct product of finite Lie algebras $\varphi(L) \subseteq A$ which proves the corollary.

The structure of $\mathfrak{V}$-free algebras is rather intricate. A beam of light onto this subject will be thrown in sections which follow.

3.2. Verbal subalgebras

By *words* we shall mean elements of (absolutely) free Lie algebras. Given a set V of words, $V \subseteq L(X)$, for some set X, and a Lie algebra A over the same ring k, denote by $V(A)$ and call a *verbal subalgebra* corresponding to V the least ideal of A containing all values of all words from V in A. Thus, for example, A^{c+1} is a verbal subalgebra of A corresponding to the word $[x_{c+1}, x_c, \dots, x_1]$, $A^{(l)}$ is a verbal subalgebra corresponding to $\delta_l(x_1, x_2, \dots, x_{2^l})$ and so on. One readily verifies that verbal subalgebras are *fully invariant*, that is closed under all endomorphisms of A. Easy but useful is the

Proposition 1. *Let $\varphi: A \to B$ be an epimorphism, V be a set of words, then $\varphi(V(A)) = V(B)$.*

An importance of verbal subalgebras is partly justified by one more

Proposition 2. *Let V be a set of words, $V \subseteq L(X)$, X being some set, $w \in L(X)$. An identical relation $w \equiv 0$ is a consequence of a set $\{v \equiv 0, v \in V\}$ of identical relations if and only if $w \in V(L(X))$.*

Proof. Assume first that $w \equiv 0$ is a consequence of the above mentioned set of identities. Put $B = L/V(L)$ and denote by ε the natural epimorphism of L onto B. Since B satisfies all the identities $v \equiv 0$, $v \in V$, $\varepsilon(w) \equiv 0$ and thus $w \in \operatorname{Ker} \varepsilon = V(L)$.

The converse inclusion is even more trivial, since $V(L)$ lies in the kernel of any homomorphism of L into any algebra satisfying $v \equiv 0$, $v \in V$.

Note also that by these two propositions, identities $v \equiv 0$, $v \in V$, hold in B if and only if $V(B) = \{0\}$.

Corollary 1. *A complete set of identities of any variety $\mathfrak{V}$ forms an ideal $V(L)$ of a free Lie algebra $L = L(X)$, X being some countable set, V be any subset of $L(X)$ such that $\mathfrak{V}$ is determined by the identities $v \equiv 0$, $v \in V$.*

Corollary 2. *There exists an antiisomorphism between the lattice of all varieties of Lie algebras over k and the lattice of all verbal subalgebras of a free Lie algebra $L(X)$, X being countable.*

Remark. Let Y be a set, $\mathfrak{B}$ be a variety, $L(Y, \mathfrak{B})$ be a $\mathfrak{B}$-free algebra with a set Y of free generators, $L(Y)$ be an absolutely free algebra. Then $L(Y, \mathfrak{B}) \cong L(Y)/V(L(Y))$.

One of the most interesting questions of the theory of Lie algebra varieties is whether any variety of Lie algebras can be determined by a finite set of identical relations. By the corollary 1 this is equivalent to the question: whether any verbal subalgebra of a free Lie algebra of countable rank can be generated as a verbal subalgebra by a finite number of words. This question has negative answer in the case of a field k of prime characteristic p; we shall see this in section 5 of this chapter. On the other hand it will be shown in the chapter 5 that the identities of any finite Lie algebra are finitely based. Probably the most intriguing question is the case of a field k of characteristic zero. This situation contrasts to the fact that in some senses the varieties of Lie algebras over fields of zero characteristic are more simple as will be seen in the two following sections.

3.3. Linearization of identities

In the theory of Lie algebra varieties an important role is played by multilinear, multihomogeneous and homogeneous identities, that is by identical relations whose left hand side is respectively multilinear, multihomogeneous or homogeneous in a free Lie algebra (see section 2, 1).

Proposition 1. *Every nontrivial identity has a nontrivial multilinear consequence.*

Proof. Let $v(x_1, x_2, \dots, x_n) \equiv 0$ be some identical relation such that $v = v(x_1, x_2, \dots, x_n)$ is not multilinear. Let $v = \sum_\alpha v_\alpha$ be the multihomogeneous decomposition of v (cf. 2, 1). Put $v_1 = \sum_{\alpha(x_1)=0} v_\alpha$, $v_2 = v - v_1$. Then $v_1 = v(0, x_2, \dots, x_n) \in v(L)$ and if $v_1 \neq 0$, the proof can be accomplished by the induction on n. Thus assume $v_1 = 0$ and then for some α, $\alpha(x_1) > 1$, $v_\alpha \neq 0$. Suppose also that $\alpha(x_1) \geqq \beta(x_i)$ for all β with $v_\beta \neq 0$. Let $y, z \in X \setminus \{x_1, x_2, \dots, x_n\}$ and consider an element from $v(L)$ having the form

$$w(y, z, x_2, \dots, x_n) = v(y + z, x_2, \dots, x_n) - v(y, x_2, \dots, x_n) - v(z, x_2, \dots, x_n) . \tag{1}$$

Using the natural embedding of $L(X)$ into $A(X)$ (section 2,2), we show that $w(y, z, x_2, \dots, x_n)$ determined by (1) is a nonzero element of $A(X)$. Indeed, since for different α the elements

$$w^\alpha = v_\alpha(y + z, x_2, \dots, x_n) - v_\alpha(y, x_2, \dots, x_n) - v_\alpha(z, x_2, \dots, x_n)$$

are linearly independent (they have different multihomogeneous decompositions) and since for any β such that $w^\alpha_\beta \neq 0$, $\beta(y)$, $\beta(z) < \beta(x_1)$ and $\beta(x_i) = \alpha(x_i)$, $i \geqq 2$, to complete the proof it suffices to show that if v is a nonzero multihomogeneous element of $A(X)$ then w determined by (1) is also nonzero. Thus let $\mu a_0 x_1^{l_1} a_1 x_1^{l_2} \dots a_{s-1} x_1^{l_s} a_s$ by some monomial from $A_\alpha(X)$, $\mu \in k$, $l_1 + l_2 + \dots + l_s = \alpha(x_1)$, which is a nonzero summand of v. Then only this monomial gives rise to the monomial $\mu a_0 y z^{l_1-1} a_1 z^{l_2} a_2 \dots \dots a_{s-1} z^{l_s} a_s$ which is a nonzero summand of $w(y, z, x_2, \dots, x_n)$.

This procedure and the induction over the number of x_i, $\beta(x_i) = \alpha(x_1)$ for some β with $v_\beta \neq 0$ complete the proof of the proposition.

In many important cases this proposition may be inverted.

Proposition 2. *Let k be an infinite field, $v \equiv 0$ be some identity. Then there exists a set $\{w \equiv 0,\ w \in W\}$ of multihomogeneous identities equivalent to $v \equiv 0$; clearly then $W \subseteq v(L)$.*

Proof. Let $v = \sum_\alpha v_\alpha$ be a multihomogeneous decomposition of v. Put $v_i = \sum_{\alpha(x_1)=i} v_\alpha$. Then $v = v_0 + v_1 + \dots + v_t$ for some $t \geqq 0$. Let $\lambda_0, \lambda_1, \dots, \lambda_t$ be pairwise different nonzero elements from k. Then $v(\lambda_i x_1, x_2, \dots, x_n)$ is an element of $v(L)$ having the form

$$v_0 + \lambda_i v_1 + \lambda_i^2 v_2 + \dots + \lambda_i^t v_t\,, \qquad i = 0, 1, 2, \dots, t\,.$$

A standard argument using Vandermonde's determinant shows that then all v_i, $i = 0, 1, \dots, t$, are themselves elements of $v(L)$. An easy induction by the number of variables in which v is not homogeneous completes the proof of the proposition.

Proposition 3. *Over a field k of characteristic zero any system of identical relations is equivalent to a system of multilinear ones.*

Proof. By the preceding proposition one can restrict oneself only to the case of a multihomogeneous identity $v(x_1, x_2, \dots, x_n) \equiv 0$. Clearly to prove the proposition it suffices to show that the identity $w \equiv 0$, w obtained from v by formula (1), is equivalent to $v \equiv 0$. In this case, of course, $\alpha(x_1) > 1$. So, one has,

$$w(x_1, x_1, x_2, \dots, x_n) = (2^{\alpha(x_1)} - 2)\, v(x_1, x_2, \dots, x_n)\,.$$

Thus, $2^{\alpha(x_1)} - 2 \neq 0$, $v \in w(L)$. The converse inclusion being trivial, our proposition is proved.

Simple examples show that the restrictions on the base ring in both propositions 2 and 3 are essential. For example, the reader may easily prove that the word $v = (\mathrm{ad}(x))^{p+1}(y)$ is a counterexample to the proposition 3 if char $k = p > 0$.

3.4. Lie algebra varieties over infinite fields

In this section we apply the above proved propositions 3.1-3 to obtain some important properties of Lie algebra varieties over infinite fields.

Proposition 1. *Let $\mathfrak{V}$ be a Lie algebra variety over an infinite field k and $A = L(Y, \mathfrak{V}) = L(Y)/V(L(Y))$ be its free algebra with some set Y of free generators, $L(Y)$ being an absolutely free Lie algebra. Then A is a graded Lie algebra* (see 2, 4), *the graduation being formed by the subspaces $A_n = L_n + V/V$, $n = 1, 2, \dots$.*

Proof. Just like in 2.1, it suffices to show that $V = V(L)$ is a homogeneous ideal of L, this being, however, an immediate corollary of proposition 3.2, or more exactly, of the proof of this latter. For, in fact, we prove in 3.2 that an element $v \in V$ if and only if its multihomogeneous components $v_\alpha \in V$.

Corollary. *If $A = L(Y, \mathfrak{V})$ is a $\mathfrak{V}$-free algebra over an infinite field k, then $\bigcap_{n=1}^{\infty} A^n = \{0\}$.*

A useful application of this proposition is the following

Proposition 2. *Given a $\mathfrak{V}$-free algebra $A = L(Y, \mathfrak{V})$, $\mathfrak{V}$ being some Lie algebra variety over some infinite field k, its subset Z linearly independent modulo A^2, and a subalgebra $B = \mathrm{alg}(Z)$ generated by Z. Then $B \cong L(Z, \mathfrak{V})$, a $\mathfrak{V}$-free algebra with the set Z of free generators.*

Proof. Let, firstly, $\{z_1, z_2, \dots, z_m\}$ be some finite subset of Z. We prove that $B^* = \mathrm{alg}\{z_1, z_2, \dots, z_m\}$ is a $\mathfrak{V}$-free algebra freely generated by $z_1, z_2, \dots, z_m$. We choose $y_1, y_2, \dots, y_n$ so that $A^* = \mathrm{alg}(\{y_1, y_2, \dots, y_n\}) \supseteq B^* = \mathrm{alg}(\{z_1, z_2, \dots, z_m\})$. The linear independence of $z_1, z_2, \dots, z_m$ implies $n \geqq m$. Since Y is linearly independent modulo A^2, one may permute $y_1, y_2, \dots, y_n$, if necessary, to get a linearly independent modulo A^2 system $\{z_1, z_2, \dots, z_m, y_{m+1}, \dots, y_n\}$. Clearly $A^* = \mathrm{alg}(\{y_1, y_2, \dots, y_n\}) \supseteq B' = \mathrm{alg}(\{z_1, z_2, \dots, z_m, y_{m+1}, \dots, y_n\})$. A^* being $\mathfrak{V}$-free, the mapping $y_1 \mapsto z_1, \dots, y_m \mapsto z_m, y_{m+1} \mapsto y_{m+1}, \dots, y_n \mapsto y_n$ extends to the monomorphism $\chi: A^* \to A^*$. For, assume $0 \neq w \in \mathrm{Ker}\, \chi$. Since, by the preceding corollary, $\bigcap\limits_{t=1}^{\infty} (A^*)^t = \{0\}$, there exists a $t > 1$ such that $w \notin (A^*)^t$, and if $\bar{\chi}: A^*/(A^*)^t \to A^*/(A^*)^t$ is a homomorphism induced by χ, then $\bar{\chi}(\bar{w}) = 0, \bar{w} = w + (A^*)^t \neq 0$. But $\bar{\chi}(A^*/(A^*)^t) \supseteq \bar{B}' + (\bar{A}^*)^t = \bar{A}^* + (\bar{A}^*)^t = A^*/(A^*)^t$, by proposition 1, 8.4, since B' contains a subset $\{z_1, \dots, z_m, y_{m+1}, \dots, y_n\}$ generating A^* modulo $(A^*)^2$. Thus $\bar{\chi}$ is an epimorphism of a finite dimensional linear space $A^*/(A^*)^t$, and therefore an isomorphism. This contradiction proves that χ is a monomorphism and, as $\chi(A^*) = B'$, this latter algebra is $\mathfrak{V}$-free with a set $\{z_1, \dots, z_m, y_{m+1}, \dots, y_n\}$ of free generators. In particular, $B^* = \mathrm{alg}\,(\{z_1, z_2, \dots, z_m\})$ is $\mathfrak{V}$-free.

Given, secondly, an infinite set X of the same cardinality as Z and a bijection $\varphi: X \to Z$, this latter extends to a homomorphism $\bar{\varphi}: L(X, \mathfrak{V}) \to B = \mathrm{alg}(Z)$. By the first part of the proof one immediately sees that $\bar{\varphi}$ is in fact an isomorphism, which completes the proof of the proposition.

Corollary. *Any system of generators of a $\mathfrak{V}$-free algebra A over some infinite field linearly independent modulo its derived subalgebra A^2 is a system of free generators of A.*

We see from the proposition that an important information about a $\mathfrak{V}$-free algebra $A = L/V(L)$ is contained in the structure of $V(L)$. The following theorem 2 gives a sort of description of such subalgebras. Recall, that by a *characteristic* subalgebra one means a subalgebra stable under all derivations of the ambient algebra; an *invariant* subalgebra B of some Lie algebra A is a subalgebra closed under all automorphisms of A.

Lemma 1. *Given an absolutely free Lie algebra $L = L(X)$ with an infinite set X of free generators over some ring k, denote by φ_x, $x \in X$, an endomorphism of $L(X)$ given by $\varphi_x(x) = 0$ and $\varphi_x(y) = y$, $y \in X \setminus \{x\}$. An invariant subalgebra $R \subseteq L$ closed under φ_x for all $x \in X$ is fully invariant.*

Proof. Assume that $w = w(x_1, x_2, \dots, x_m)$ is an element of R and that φ is an endomorphism of L such that $\varphi(x_i) = v_i(x_1, x_2, \dots, x_n)$, $i = 1, 2, \dots, m$. Consider an automorphism ω of $L(X)$ given by $\omega(x_i) = x_{n+1} + v_i(x_1, x_2, \dots, x_n)$, $1 \leqq i \leqq m$; $\omega(x_{m+j}) = x_j$, $1 \leqq j \leqq n$; $\omega(y) = y$, $y \in X \setminus \{x_1, x_2, \dots, x_{m+n}\}$. The bijectivity of ω immediately follows from the corollary to the proposition 2 of this section. By our conditions $\omega(w(x_1, x_2, \dots, x_m)) = w(x_{n+1} + v_1(x_1, \dots, x_n), x_{n+2} + v_2(x_1, \dots, x_n), \dots, x_{n+m} +$

$v_m(x_1, \dots x_n))$ is an element of R. But then, also by the conditions of the lemma $\varphi_{x_{n+1}}\varphi_{x_{n+2}} \cdots \varphi_{x_{n+m}}(\omega(w(x_1, x_2, \dots, x_m))) \in R$. This element, however, equals $w(v_1(x_1, \dots, x_n), v_2(x_1, \dots, x_n), \dots, v_m(x_1, \dots, x_n))$ and the lemma is proved.

Now we formulate the theorem itself.

Theorem 2. *Let $L = L(X)$ be a free Lie algebra over an infinite field k with an infinite set X of free generators, and R be its invariant subalgebra. Then R is a verbal subalgebra $V(L)$ for some set V of words.*

Proof. We show, first, that R is fully invariant. By the preceding lemma it suffices to verify that $\varphi_x(R) \subseteq R$ for all $x \in X$. Represent u as a sum of elements $u_0 + u_1 + \dots + u_s$, $u_i = \sum_{\alpha_i(x)=i} u_{\alpha_i}$. Use induction by s, the case $s = 0$ being trivial. Let then $s > 0$ and u do not depend on $z \in X$. Consider the automorphisms θ_1 and θ_2 of L induced by $x \mapsto z$, $z \mapsto x$, and $x \mapsto x + z$, $z \mapsto z$ respectively $(\theta_i(y) = y,\ y \neq x, z,\ i = 1, 2)$. By our assumptions $v = u + \theta_1(u) - \theta_2(u) \in R$. But like in the proof of the proposition 3.1, one sees that the degree of v in x is strictly less than that of u, and by the induction hypothesis, using $\varphi_x(\theta_2(u)) = \theta_1(u)$, one has $R \ni \varphi_x(v) = \varphi_x(u) + \theta_1(u) - \theta_1(u) = \varphi_x(u)$. This completes the proof of the fact that R is fully invariant.

We prove, second, that R is an ideal of L. By the proof of the proposition 3.1, R is a multihomogeneous subalgebra of L. Thus it suffices to prove that if $v(x_1, x_2, \dots, x_n)$ is a multihomogeneous element from R and $x_{n+1} \in X \setminus \{x_1, \dots, x_n\}$ then $[v(x_1, x_2, \dots, x_n), x_{n+1}] \in R$. But this element is a sum of multihomogeneous parts w_α of the element

$$w = v(x_1 + [x_1, x_{n+1}], x_2 + [x_2, x_{n+1}], \dots, x_n + [x_n, x_{n+1}]),$$

with $\alpha(x_{n+1}) = 1$. Since $w \in R$, the element $[v(x_1, x_2, \dots, x_n), x_{n+1}] \in R$. Given now an arbitrary element $a \in A$, consider a homomorphism $\varphi: A \to A$ extending the map $x_i \mapsto x_i, i \neq n + 1; x_{n+1} \mapsto a$. Then $[v(x_1, x_2, \dots, x_n), a] = \varphi([v(x_1, x_2, \dots, x_n), x_{n+1}]) \in R$ and so R is an ideal of A. Thus the theorem is proved, for it suffices to put $V = R$.

Corollary. *Let k be a field of characteristic zero, $L(X)$ be a free Lie algebra over k with an infinite set X of free generators. Then any invariant subalgebra R of L is a characteristic verbal subalgebra of L.*

It rests to show that if $v(x_1, x_2, \dots, x_n)$ is a multilinear element of R (see proposition 3.3) and d is a derivation of L then $d(v(x_1, x_2, \dots, x_n)) \in R$. But this immediately follows from the formula below valid for the multilinear words

$$d(v(x_1, x_2, \dots, x_n)) = \sum_{i=1}^{n} v(x_1, \dots, d(x_i), \dots, x_n).$$

3.5. Product varieties

Let $\mathfrak{U}$, $\mathfrak{B}$ be two varieties of Lie algebras over the same ring k. Denote by $\mathfrak{U}\mathfrak{B}$ the class of all Lie algebras A possessing an ideal $B \in \mathfrak{U}$ such that $A/B \in \mathfrak{B}$. By the characterization of Lie algebra varieties given in the theorem 1.1, $\mathfrak{U}\mathfrak{B}$ is a variety called a *product* of $\mathfrak{U}$ by $\mathfrak{B}$. Clearly, $\mathfrak{U}\mathfrak{B} \supseteq \mathfrak{U}, \mathfrak{B}$, and we call a variety $\mathfrak{W}$ an *indecomposable* one if it is not a product of its proper subvarieties. Denoting by $\mathfrak{D}$ the variety of all

Lie algebras over k and by $\mathfrak{E}$ the one element variety (that is containing only the zero algebra), one immediately sees that these varieties play roles of respectively zero and neutral elements in the groupoid $v(k)$ of all Lie algebra varieties over k.

Lemma 1. *Assume that k is an infinite field. Then $v(k)$ is a semigroup.*

Proof. Given an absolutely free Lie algebra L of countable rank and varieties $\mathfrak{U}, \mathfrak{V}, \mathfrak{W} \in v(k)$ determined by sets U, V, W of words, by repeated application of theorem 4.2 one sees that $R = U\big(V(W(L))\big)$ is a verbal subalgebra of L. Then if φ is a homomorphism of L onto some algebra A from, say, $(\mathfrak{U}\mathfrak{V})\,\mathfrak{W}$, then by proposition 2.1 $\varphi(R) = U\Big(\varphi\big(V(W(L))\big)\Big) = U\Big(V\big(\varphi(W(L))\big)\Big) = U\Big(V\big(W(\varphi(L))\big)\Big) = U\big(V(W(A))\big) = \{0\}$; for, if B is an ideal of A belonging to $\mathfrak{U}\mathfrak{V}$ with $A/B \in \mathfrak{W}$ then $W(A) \subseteq B$, $U\big(V(B)\big) = \{0\}$. Similarly, R is contained in the kernel of any epimorphism of L onto some algebra from $\mathfrak{U}(\mathfrak{V}\mathfrak{W})$. Since further $F = L/U\big(V(W(L))\big)$ belongs to both of these varieties, it is a free algebra of both of them, that is $(\mathfrak{U}\mathfrak{V})\,\mathfrak{W} = \mathfrak{U}(\mathfrak{V}\mathfrak{W})$ and $U\big(V(W(L))\big)$ is the set of all words determining the variety $\mathfrak{U}\mathfrak{V}\mathfrak{W}$.

The top result of this section will be theorem 4 which states that in fact $v(k)$, k being an infinite field, is a free semigroup with 1 and 0. But to prove this we need some preliminary results some of them seem to be of their own interest. The first of them is the

Theorem 3 (BAHTURIN). *Let L be a nonabelian absolutely free Lie algebra over some field k, $\mathfrak{V}$ be a nonzero variety determined by a set V of words, R, S be ideals of L. Then $V(R) \subseteq V(S)$ implies $R \subseteq S$.*

Proof. Let the conditions of the theorem be satisfied, but $R \not\subseteq S$, so that there exists an element $r \in R \setminus S$. Consider $R_1 = R \cap S + kr$, $L_1 = R_1 + S$, L_1 thus being a nonabelian free Lie algebra, R_1, S being its ideals such that $V(R_1) \subseteq V(R) \subseteq V(S)$; but $R_1 \not\subseteq S$. Thus in the proof we may restrict ourselves to the case $L = R + S$ and $\dim(L/S) = 1$ (for $L_1/S \cong R_1 + S/S \cong R_1/R_1 \cap S = R_1/R \cap S$). Using the proof of the corollary of theorem 2, 3.4, there exists a system $Z = \{x\} \cup Z'$ of free generators of L such that $S = \mathrm{id}_L(Z')$ and $x \notin S$. Choose then an element $y \in Z'$ and consider an epimorphism $\varphi\colon L(Z) \to L(\{x, y\})$ given by $\varphi(x) = x$, $\varphi(y) = y$, $\varphi(Z') = \{0\}$. Under this homomorphism $L = L(Z)$ is mapped onto its free subalgebra $L_1 = L(\{x, y\})$, whence, denoting $R_1 = \varphi(R)$, $S_1 = \varphi(S)$, one has $L_1 = R_1 + S_1$, $\dim\,(L_1, S_1) = 1$, $V(R_1) = V\big(\varphi(R)\big) = \varphi\big(V(R)\big) \subseteq \varphi\big(V(S)\big) \subseteq V(S_1)$, but $R_1 \not\subseteq S_1$. Therefore, the only case it rests to deal with is that of a free Lie algebra L freely generated by $\{x, y\}$ and $L = R + S$, $S = \mathrm{id}_L(\{y\})$.

Denote then by $N(\mathfrak{V})$ the *norm* of the variety $\mathfrak{V}$, that is the least degree of junior components (see 2,5.5) of words v such that $v \equiv 0$ is an identical relation in $\mathfrak{V}$. One immediately deduces from proposition 3.1 that if d is the norm of $\mathfrak{V}$ then there exists a word v with multilinear junior part w of degree d such that $v \equiv 0$ is an identity in $\mathfrak{V}$. Choose such a word $v = v(z_0, z_1, \dots, z_{d-1})$ with the multilinear junior part $w(z_0, z_1, \dots, z_{d-1})$.

Now using lemma 2,4.2, one observes that S is a free Lie algebra freely generated by a set T of monomials t_q, having form $t_q = \big(\mathrm{ad}(x)\big)^q\,(y)$, $q = 0, 1, 2, \dots$. We order T by $t_i < t_j$ if and only if $i < j$, and construct some system of basis monomials in T whose total order extends the ordering of T. We construct also a system of basic monomials in

$\{z_0, z_1, \dots, z_{d-1}\}$ induced by the embedding ι of $L(\{z_0, z_1, \dots, z_{d-1}\}) \to L(T)$ extending the mapping $z_i \mapsto t_{2i}$, $i = 0, 1, \dots, d-1$.

Since, further, $y \in S$, there exists an element $r \in R \setminus S$ having the form $r = x + r_1$, where the junior part of r_1 has degree > 1 in $\{x, y\}$. Consider then a value of v in R which by our assumption also is an element of $V(S)$:

$$v\big(r, (\mathrm{ad}(t_1))\,(r), (\mathrm{ad}(t_3))\,(r), \dots, (\mathrm{ad}(t_{2d-3}))\,(r)\big)\,. \tag{1}$$

The junior part of (1) both in $\{x, y\}$ and in T has the form

$$w\big(x, (\mathrm{ad}(t_1))\,(x), (\mathrm{ad}(t_3))(x), \dots, (\mathrm{ad}(t_{2d-3}))(x)\big) = \tag{2}$$
$$= \pm\, w(x, t_2, t_4, \dots, t_{2d-2})\,.$$

Being an element of $V(S) = V\big(L(T)\big)$, (1) is an element of the ideal of $L(T)$ generated by the elements $u(t_1, t_2, \dots, t_p)$, u being some word such that $u \equiv 0$ is an identity in $\mathfrak{B}$. By the definition of the norm then the degree of the junior part (2) of (1) in T is at least d. But any homogeneous element of degree $\geqq d$ in T has degree $\geqq d$ in y which is not true for (2).

It remains therefore to prove that (2) is a nonzero element of $L(T)$. To do this represent $w(z_0, z_1, \dots, z_{d-1})$ as a linear combination of basic monomials and let $[w_r, \dots, w_3, w_2, z_0, z_i]$ be one of these monomials, $w_r > \dots > w_3 > w_2 > z_0 < z_i$ (recall that w is multilinear), $w_r, \dots, w_2$ do not depend on z_0. Then (2) is a linear combination with the same coefficients of monomials of the form $\pm\, [\iota(w_r), \dots, \iota(w_3), \iota(w_2), x, t_{2i}]$. By the definition of ι and t_q this latter monomial is a product of basic monomials and has the form

$$\pm\, \big[\iota(w_r), \dots, \iota(w_3), \iota(w), t_{2i+1}\big]\,, \tag{3}$$

$\iota(w_r) > \dots > \iota(w_3) > \iota(w_2)$. If also $\iota(w_2) < t_{2i+1}$ then (3) is basic, if $t_{2i+1} < \iota(w_2) < \iota(w_3)$ then the inverse of (3) is basic. The case $t_{2i+1} = \iota(w_2)$ being impossible by our choice in (1) and an evident distinction of basic monomials arising from different monomials of the form (3), prove that (2) is nonzero and complete the proof of the theorem.

Corollary. *Let the base ring k be some infinite field. Given a nonzero ideal R of a free Lie algebra L of infinite rank and two verbal subalgebras U, V of another free Lie algebra L_∞ of countable rank, then $U(L) = V(R)$ implies either $U = V = \{0\}$ or that there exists a verbal subalgebra $W \subseteq L_\infty$ such that $R = W(L)$ and $U = V(W)$.*

Proof. Obviously, $U \subseteq V$. If $V = \{0\}$ then $U = \{0\}$ and we are finished. Therefore assume $V \neq \{0\}$ and let φ be an automorphism of L; then $V\big(\varphi(R)\big) = \varphi\big(V(R)\big) = \varphi\big(U(L)\big) = U\big(\varphi(L)\big) = U(L) = V(R)$. By the theorem then $\varphi(R) = R$, that is R is an invariant ideal of L and hence by theorem 4.3 R is a verbal subalgebra $W(L)$ determined by some verbal subalgebra $W \subseteq L_\infty$. Now $U(L) = V\big(W(L)\big)$ and since L is of infinite rank, $U = V(W)$, which completes the corollary.

One more assertion of its own interest is the

Proposition 1 (Genov). *Let R, S be ideals of a free Lie algebra L of infinite rank over some field k, U, V be two verbal subalgebras of a free Lie algebra L_∞ of countable rank over k. Then $U(R) \subseteq V(S)$ implies either $U \subseteq V$ or $R \subseteq S$.*

Proof. Just like in the first reduction in the proof of theorem 3 one may assume that if $R \nsubseteq S$ then $L = R + S = L(Z)$, $\dim(L/S) = 1$, $Z = \{x\} \cup Z'$, $S = \mathrm{id}_L(Z')$. Since S is an ideal of a free Lie algebra of infinite rank, Z is an infinite set. Choose an element $r \in R \setminus S$ having the form $r = x + s$, $s \in S$, and let Z_1 be a finite subset of Z' such that $r \in \operatorname{alg}(\{x\} \cup Z_1)$, $Z_2 = Z' \setminus Z_1$. Consider a system P of elements of the form

$$P = \{ (\operatorname{ad}(x))^t (z), (\operatorname{ad}(r))^u (z') \mid z \in Z_1, z' \in Z_2, t, u = 0, 1, 2, \ldots\} . \tag{4}$$

We show that (4) is a system of free generators of S. To do this remark first that by lemma 2, 3.2 S is a free Lie algebra freely generated by the system

$$Q = \{ (\operatorname{ad}(x))^t (z), (\operatorname{ad}(x))^u (z') \mid z \in Z_1, z' \in Z_2, t, u = 0, 1, 2, \ldots \} . \tag{5}$$

Clearly (4) and (5) coincide elementwise modulo S^2 so that by theorem 4.2, valid for the absolutely free Lie algebra case over any field, P is the set of free generators of a subalgebra $S_1 = \operatorname{alg}(P)$. We prove that $S_1 = S$. Using induction by u we prove that $(\operatorname{ad}(x))^u (z')$ is an element of S_1. Indeed, consider

$$(\operatorname{ad}(r))^u (z') = (\operatorname{ad}(x + s))^u (z') = (\operatorname{ad}(x))^u (z') + \\ + \sum_{m_1+\ldots+m_p>0} (\operatorname{ad}(x))^{l_1} (\operatorname{ad}(s))^{m_1} \ldots (\operatorname{ad}(x))^{l_p} (\operatorname{ad}(s))^{m_p} (z') . \tag{6}$$

Using the facts that $s \in \operatorname{alg}(\{x\} \cup Z_1)$ and thus lies in a subalgebra of S_1 generated by $(\operatorname{ad}(x))^t (z)$, $z \in Z_1$, $t = 0, 1, 2, \ldots$, that Z_2 itself lies in P, and the derivation rules, one easily observes that the sum in (6) is an element of a subalgebra of S generated by $(\operatorname{ad}(x))^l (z')$ and $(\operatorname{ad}(x))^t (z)$, $0 \leqq l < u$, $t \geqq 0$, $z' \in Z_2$, $z \in Z_1$. By the induction and from (6) $(\operatorname{ad}(x))^u (z') \in S_1$, which proves thus $S = L(P)$.

Now consider $S' = \operatorname{alg}(\{(\operatorname{ad}(r))^u (z') \mid z' \in Z_2, u = 1, 2, \ldots\}) \subseteq R$. Since S' is generated by a part, say, P' of free generators P of S, $V(S') = V(S) \cap S'$ (for, clearly, $V(S') \subseteq V(S)$; let $x \in V(S) \cap S'$ and $\pi: S \to S'$ be a homomorphism induced by the projection $P \to P'$, then $x = \pi(x) \in \pi(V(S)) = V(\pi(S)) = V(S')$). So far, write $U(S') \subseteq \subseteq U(R) \cap S' \subseteq V(S) \cap S' = V(S')$. But as S' is a free Lie algebra of infinite rank, one has $U \subseteq V$, and the proof is complete.

Corollary. *Let R, S be two nonzero ideals of a free Lie algebra L of infinite rank over some infinite field k, U, V be two verbal subalgebras of some free Lie algebra L_∞ of countable rank over k with $U(R) = V(S)$. Then there exists a verbal subalgebra $W \subseteq L_\infty$ such that either $R = W(S)$ and $U(W) = V$ or $S = W(R)$ and $U = V(W)$.*

Proof. From the condition $U(R) = V(S)$ and proposition 1 one immediately deduces that either $U = V$ and then by theorem 3, $R = S$, or one of the two ideals R, S is contained in the other one. If, say, $R \subseteq S$, then usingthe corollary to theorem 3 and the above mentioned fact that S is a free Lie algebra of infinite rank one easily obtains the desired.

Now we are in a position to prove the main in this section:

Theorem 4 (Parfenov). *Let $v(k)$ be a semigroup of varieties over some infinite field k. Then $v(k)$ is a free semigroup with zero $\mathfrak{D}$ and neutral element $\mathfrak{E}$ freely generated by indecomposable varieties.*

Proof. By a *nontrivial* variety we shall mean a variety $\mathfrak{V} \neq \mathfrak{D}, \mathfrak{E}$.

To show that any nontrivial variety is a product of indecomposable ones it suffices to show that given nontrivial varieties $\mathfrak{U}$, $\mathfrak{V}$, $\mathfrak{W}$ with $\mathfrak{W} = \mathfrak{U}\mathfrak{V}$, then $N(\mathfrak{W}) \geqq \geqq N(\mathfrak{U})\, N(\mathfrak{W})$. To do this, note that if U, V, W be verbal subalgebras determining $\mathfrak{U}$, $\mathfrak{V}$, $\mathfrak{W}$ respectively then $W = U(V)$, the argument being the same as in lemma 1.

Now having in mind that by proposition 3.2 U, V and W are homogeneous and thus for example $N(\mathfrak{U})$ is the least degree of (homogeneous and by proposition 3.1 even multilinear) elements from U, one readily obtains the above written inequality for norms (in fact the reader may as well easily obtain $N(\mathfrak{W}) = N(\mathfrak{U})\, N(\mathfrak{V})$). The trivial remark that $N(\mathfrak{U}) = 1$ implies $\mathfrak{U} = \mathfrak{E}$ completes the proof of the statement that every nontrivial variety is a product of indecomposable nontrivial ones.

Assume further that $\mathfrak{U}_1, \mathfrak{U}_2, \dots, \mathfrak{U}_s, \mathfrak{V}_1, \mathfrak{V}_2, \dots, \mathfrak{V}_r$ are indecomposable varieties such that

$$\mathfrak{U}_1\mathfrak{U}_2 \dots \mathfrak{U}_s = \mathfrak{V}_1\mathfrak{V}_2 \dots \mathfrak{V}_r . \tag{7}$$

Denoting $\mathfrak{U}_2 \dots \mathfrak{U}_s$ by $\mathfrak{U}_1'$ and $\mathfrak{V}_2 \dots \mathfrak{V}_r$ by $\mathfrak{V}_1'$ one rewrites (7) in the form $\mathfrak{U}_1\mathfrak{U}_1' = \mathfrak{V}_1\mathfrak{V}_1'$, or $U_1(U_1') = V_1(V_1')$, where U_1, U_1', V_1, V_1' be the corresponding verbal subalgebras of L_∞. By the preceding corollary there exists a $W \subseteq L_\infty$ such that either $U_1 = V_1(W)$ or $V_1 = U_1(W)$. Since $\mathfrak{U}_1$ and $\mathfrak{V}_1$ are indecomposable, $W = L_\infty$ and $U_1 = V_1$. But in this case by the same corollary $U_1' = V_1'$ and the proof of the uniqueness of the decomposition (7) is completed using induction by s.

3.6. Some infinitely based varieties of Lie algebras

In spite of the title of the section we start with an argument which shows that any nilpotent variety of Lie algebras over any noetherian ring k is finitely based.

Let $\mathfrak{V}$ be a nilpotent variety, L_∞ be an absolutely free Lie algebra of countable rank, V be its verbal subalgebra determining $\mathfrak{V}$. Then for some $c \geqq 1$, $V \supseteq L^{c+1}$. We prove that modulo L^c any identical relation of the form $w \equiv 0$ is equivalent to a set of identities $\{v_\alpha \equiv 0\}$ whose left hand sides depend on at most c variables. Let $w = w(x_1, x_2, \dots, x_n)$ depend on $n > c$ variables; we proceed by induction on n. Denote by δ_i, $i = 1, 2, \dots, n$, the homomorphisms of L_∞ extending the maps $\delta_i(x_j) = x_j$, $j \neq i$, and $\delta_i(x_i) = 0$. Put $u = (1 - \delta_1)(1 - \delta_2) \dots (1 - \delta_n)(w)$; then $u \in \operatorname{alg}(\{x_1, x_2, \dots, x_n\})$ on one hand and $\delta_i(u) = 0$ on the other hand. This means that $u \in L^n \subseteq L^{c+1}$ and therefore modulo L^c w is representable as a sum of elements $\beta(w)$, where β is a nonempty product of *deletions* δ_i, that is $\beta(w)$ depends on less than n variables. Since each $\beta(w)$ is a consequence of w, our assertion follows.

Let L_1 be a subalgebra of L_∞ freely generated by a subset $\{x_1, x_2, \dots, x_{c+1}\}$ of a set of free generators of L_∞. As we have seen just above, V is generated as a verbal subalgebra by $V \cap L(\{x_1, x_2, \dots, x_c\})$ and $[x_1, x_2, \dots, x_{c+1}]$, hence by $V \cap L_1 = V(L_1)$. Since $V(L_1)/L_1^c$ is a submodule of a k-module L_1/L_1^c of finite type (see proposition 1, 7.2) and k is noetherian, $V(L_1)/L_1^c$ is a finite type k-module and therefore $V(L_1)$ is finitely generated as a verbal subalgebra by a finite number of the k-module generators of $V(L_1)$ and $[x_1, x_2, \dots, x_{c+1}]$. Thus we have proved

Proposition 1. *Let k be a noetherian ring, $\mathfrak{V}$ be a nilpotent Lie algebra variety over k. Then $\mathfrak{V}$ has a finite basis for its identical relations.*

We shall have in the future some more classes of finitely based varieties, but now we turn to examples of Lie algebra varieties having no finite bases for their identical relations.

Theorem 5 (Drenski). *Let k be a field of positive characteristic $p > 0$. There exists a subvariety $\mathfrak{B}$ of a variety $\mathfrak{A}^3 \cap \mathfrak{N}_p\mathfrak{A}$ having no finite basis for its identical relations.*

We need some auxiliary lemmas and constructions.

Denote by T_n, for each positive n, the set of all subsets of the set $\{1, 2, \dots, 2n-1\}$. Given $\tau \in T_n$, by $|\tau|$ we mean the cardinality of τ. Construct an algebra A_n over k with linear basis

$$\{a_\sigma^{(1)}, a_\tau^{(s)}, b \mid \sigma, \tau \in T_n, |\tau| \text{ is odd}, s = 2, 3, \dots, p-1\} \tag{1}$$

and a multiplication table

$$[a_\sigma^{(s)}, a_\tau^{(r)}] = \begin{cases} (-1)^{|\sigma|}\ a_\pi^{(s+r)}, & \pi = \sigma \cup \tau, \text{ if } s + r \leqq p, \sigma \cap \tau = \emptyset, \\ & |\sigma| + |\tau| \equiv 1 \pmod 2 \\ 0, & \text{otherwise.} \end{cases} \tag{2}$$

Here for the conveniency sake we put $a_\tau^{(p)} = 0$ if $|\tau| < 2n-1$ and $a_\tau^{(p)} = b$, if $|\tau| = 2n-1$.

Lemma 1. *A_n is a nilpotent Lie algebra of nilpotency class p and solvable length 2.*

For, if (2) is nonzero, then $|\sigma| + |\tau| \equiv 1 \pmod 2$, so that $(-1)^{|\sigma|} = -(-1)^{|\tau|}$ and the anticommutativity follows. If further

$$[a_\varrho^{(r)}, [a_\sigma^{(s)}, a_\tau^{(t)}]] + [a_\sigma^{(s)}, [a_\tau^{(t)}, a_\varrho^{(r)}]] + [a_\tau^{(t)}, [a_\varrho^{(r)}, a_\sigma^{(s)}]] \neq 0 \tag{3}$$

then by (2) one of the numbers $|\varrho|, |\sigma|, |\tau|$, say $|\varrho|$, is odd and the two others are even; moreover $\varrho \cap \sigma = \sigma \cap \tau = \varrho \cap \tau = \emptyset$ and $r + s + t \leqq p$. But then using (2) the left hand side of (3) has the form $0 + a_{\sigma\cup\tau\cup\varrho}^{(s+t+r)} - a_{\tau\cup\varrho\cup\sigma}^{(t+r+s)} = 0$ which is a contradiction. Thus the Jacobi identity holds and the lemma is proved, the assertions on the nilpotency class and the solvable length following immediately from (2).

Consider then a set

$$\{g_\lambda, h_\mu \mid \lambda, \mu \in T_n, |\lambda| \text{ is odd}, [\mu] \text{ is even}\} \tag{4}$$

of linear transformations of A_n defined by the following rules

$$(g_\lambda)\ (a_\sigma^{(s)}) = \begin{cases} a_\tau^{(s)}, & \text{where } \tau = \sigma \cup \lambda, \text{ if } \sigma \cap \lambda = \emptyset \text{ and } |\sigma| \text{ is even} \\ 0, & \text{otherwise,} \end{cases} \tag{5}$$

$$(h_\mu)\ (a_\sigma^{(s)}) = \begin{cases} s a_\tau^{(s)}, & \text{where } \tau = \sigma \cup \mu, \text{ if } \sigma \cap \mu = \emptyset \\ 0, & \text{otherwise.} \end{cases} \tag{6}$$

Lemma 2. *The set (4) is a set of pairwise commuting derivations of A_n.*

Indeed, it suffices to check up the following equations

$$g_\lambda([a_\sigma^{(s)}, a_\tau^{(t)}]) = [g_\lambda(a_\sigma^{(s)}), a_\tau^{(t)}] + [a_\sigma^{(s)}, g_\lambda(a_\tau^{(t)})], \tag{7}$$

$$h_\mu([a_\sigma^{(s)}, a_\tau^{(t)}]) = [h_\mu(a_\sigma^{(s)}), a_\tau^{(t)}] + [a_\sigma^{(s)}, h_\mu(a_\tau^{(t)})]. \tag{8}$$

But the left hand side of (7) equals zero, since either $[a_\sigma^{(s)}, a_\tau^{(t)}] = 0$ or $[a_\sigma^{(s)}, a_\tau^{(t)}] = (-1)^{|\sigma|} a_{\sigma\cup\tau}^{(s+t)}$, where $|\sigma \cup \tau|$ is odd. If the right hand side of (7), denoted by x, is nonzero then $\lambda \cap \sigma = \sigma \cap \tau = \tau \cap \lambda = \emptyset$ and at least one of $|\sigma|, |\tau|$ is even. If, say $|\sigma|$, is even and $|\tau|$ is odd, $x = [a_{\sigma\cup\lambda}^{(s)}, a_\tau^{(t)}] = 0$ since $|\sigma \cup \lambda \cup \tau|$ is odd; if besides $|\tau|$ is even, $x = [a_{\sigma\cup\lambda}^{(s)}, a_\tau^{(t)}] + [a_\sigma^{(s)}, a_{\tau\cup\lambda}^{(t)}] = -a_{\sigma\cup\lambda\cup\tau}^{(s+t)} + a_{\sigma\cup\tau\cup\lambda}^{(s+t)} = 0$ (remark that $|\sigma \cup \lambda|$ is odd and $|\sigma|$ is even).

The other equation (8) is dealt with similarly. The commutativity of (4) immediately follows from (5) and (6), and the proof is complete.

Denote then by D_n the abelian subalgebra of $\mathrm{Der}(A_n)$ spanned by (4), and let $B_n = D_n \curlywedge A_n$, the semidirect product of D_n by A_n with natural action of D_n in A_n (see 1,5).

Lemma 3. *For each $n \geqq 1$, $B_n \in \mathfrak{A}^3 \cap \mathfrak{N}_p\mathfrak{A}$; moreover,*

$$[x_{2p+1}, [x_{2p}, x_{2p-1}], [x_{2p-2}, x_{2p-3}], \dots, [x_2, x_1]] \equiv 0 \tag{9}$$

holds in B_n.

Clearly, $B_n^2 \subseteq A_n$; since by lemma 1, A_n is metabelian (that is of solvable length 2), B_n has solvable length 3, or $B_n \in \mathfrak{A}^3 = \mathfrak{S}_3$. Now by (2), (5) and (6), $b = a_\sigma^{(p)}$, $|\sigma| = 2n - 1$, is central in B_n; besides, $A_n^p \subseteq kb$, whence (9) holds and thus $B_n \in \mathfrak{N}_p\mathfrak{A}$.

More difficult is the next

Lemma 4. *Let $a, a_1, a_2 \in A_n$; $\alpha_1, \alpha_2 \in k$; $\lambda, \mu \in T_n$, $|\lambda|$ being odd, $|\mu|$ being even. Then $(\mathrm{ad}(a))^p (h\mu) = 0$ and*

$$(\mathrm{ad}(\alpha_1 a_1 + \alpha_2 a_2))^p (g_\lambda) = \alpha_1^p(\mathrm{ad}(a_1))^p (g_\lambda) + \alpha_2^p(\mathrm{ad}(a_2))^p (g_\lambda) . \tag{10}$$

Proof. Let L and M be the linear subspaces of A spanned by $\{a_\sigma^{(1)} \mid \sigma \in T_n\}$ and $\{a_\tau^{(t)} \mid \tau \in T_n, |\tau| \text{ is odd}, t = 2, 3, \dots, p\}$ respectively. Then clearly, $A_n = L \oplus M$, $A_n^2 = M$, $M^2 = \{0\}$. Given further $l \in L, m \in M$, one obtains using $A_n^{p+1} = \{0\}$, that $(\mathrm{ad}(m))^p (g_\lambda) = (\mathrm{ad}(m))^p (h_\mu) = 0$ and

$$(\mathrm{ad}(l + m))^p (g_\lambda) = (\mathrm{ad}(l))^p (g_\lambda) , \qquad (\mathrm{ad}(l + m))^p (h_\mu) = (\mathrm{ad}(l))^p (h_\mu) ,$$

so it suffices to limit ourselves to the case $a, a_1, a_2 \in L$.

Thus let $a = \sum_\sigma \alpha_\sigma a_\sigma^{(1)}$, $\alpha_\sigma \in k$, $\sigma \in T_n$. Then

$$(\mathrm{ad}(a))^2 (h_\mu) = [(\mathrm{ad}(h_\mu)) (a), a] = \sum_{\sigma,\tau} \alpha_\sigma\alpha_\tau[(\mathrm{ad}(h_\mu)) (a_\sigma^{(1)}), a_\tau^{(1)}] = 0 ,$$

for $[h_\mu(a_\sigma^{(1)}), a_\sigma^{(1)}] = 0$ since $|\mu|$ is even and $|\sigma \cup \mu| \equiv |\sigma| \pmod 2$; $[h_\mu(a_\sigma^{(1)}), a_\tau^{(1)}] + [a_\sigma^{(1)}, h_\mu(a_\tau^{(1)})] = 0$ by the same reason if $|\sigma| \equiv |\tau| \pmod 2$ and by the multiplication rule (2) otherwise. Therefore we have proved the first equality of the lemma.

Remark now, that in any metabelian algebra the following identity holds:

$$[x, y, z, u] \equiv [y, x, z, u] . \tag{11}$$

For, by the Jacobi identity, $[x, y, z, u] = [y, x, z, u] + [[x, y], [z, u]]$. The identity (11) implies evidently an identity

$$[x_1, x_2, \dots, x_t, x, y] \equiv [x_{i_1}, x_{i_2}, \dots, x_{i_t}, x, y] ,$$

where $\{i_1, i_2, \dots, i_t\} = \{1, 2, \dots, t\}$.

We show that for any $a_1, a_2, \dots, a_p \in A_n$,

$$[a_1, a_2, \dots, a_p, g_\lambda] = [a_{i_1}, a_{i_2}, \dots, a_{i_p}, g_\lambda] . \tag{12}$$

The indices $1, 2, \dots, p - 2$ are permutable in the left hand side of (12), since $[a_p, g] = -g_\lambda(a_p) \in A_n$ and A_n is metabelian. Further since $g_\lambda(M) = \{0\}$, $0 = g_\lambda([a_1, a_2]) = [g_\lambda(a_1), a_2] + [a_1, g_\lambda(a_2)] = [a_2, a_1, g_\lambda] - [a_1, a_2, g_\lambda]$, or $[a_2, a_1, g_\lambda] = [a_1, a_2, g_\lambda]$, so that p and $p - 1$ in (12) are permutable. Finally, by direct calculation $[a_\varrho^{(1)}, a_\sigma^{(1)}, a_\tau^{(1)}, g_\lambda] = [a_\sigma^{(1)}, a_\varrho^{(1)}, a_\tau^{(1)}, g_\lambda]$, whence $p - 2$ and $p - 1$ are permutable in the left hand side of (12) and therefore (12) holds. From this equality using the binomial decomposition we get

$$\left(\operatorname{ad}(\alpha_1 a_1 + \alpha_2 a_2)\right)^p (g_\lambda) = \sum_i \binom{p}{i} \alpha_1^i \alpha_2^{p-i} \left(\operatorname{ad}(a_1)\right)^i \left(\operatorname{ad}(a_2)\right)^{p-i} (g_\lambda) =$$
$$= \alpha_1^p \left(\operatorname{ad}(a_1)\right)^p (g_\lambda) + \alpha_2^p \left(\operatorname{ad}(a_2)\right)^p (g_\lambda) \,,$$

since $\binom{p}{i} \equiv 0 \pmod p$ if $i \neq 0, p$.

Thus the proof of lemma 4 is complete.

Denote now by C_n the subalgebra of B_n generated by

$$T = \{a_\emptyset^{(1)}, g_{\{1\}}, h_\emptyset, h_{\{2,3\}}, h_{\{4,5\}}, \dots, h_{\{2n-2,2n-1\}}\} \,. \tag{13}$$

Lemma 5. *If* $a_1, a_2 \in A_n$, $c_1, c_2, \dots, c_s \in C_n$; $\alpha_1, \alpha_2 \in k$, *then*

$$\left(\operatorname{ad}(\alpha_1 a_1 + \alpha_2 a_2)\right)^{p-1} ([c_1, c_2, \dots, c_s, \alpha_1 a_1 + \alpha_2 a_2]) =$$
$$= \alpha_1^p \left(\operatorname{ad}(a_1)\right)^{p-1} ([c_1, c_2, \dots, c_s, a_1]) + \alpha_2^p \left(\operatorname{ad}(a_2)\right)^{p-1} ([c_1, c_2, \dots, c_s, a_2]) \,. \tag{14}$$

Proof. Since both sides of (14) are linear in $c_1, c_2, \dots, c_s$ and as $A_n^{p+1} = \{0\}$, it suffices to consider only the case when $c_1, c_2, \dots, c_s$ are some derivations from (13). Clearly also a_1, a_2 may be taken from L (see the preceding lemma 4). Moreover since the derivations commute and $g_\lambda^2(A_n) = \{0\}$, only two cases may be considered:

(i) $c_1 = g_{\{1\}}, c_2 = h_{\mu_2}, \dots, c_s = h_{\mu_s}$, and $[c_1, c_2, \dots, c_s, a] = \pm \left(\operatorname{ad}(g_\mu)\right)(a)$, $\mu = \{1\} \cup \mu_2 \cup \dots \cup \mu_s$ or zero, for any $a \in L$;

(ii) $c_1 = h_{\mu_1}, c_2 = h_{\mu_2}, \dots, c_s = h_{\mu_s}$, and $[c_1, c_2, \dots, c_s, a] = \pm \left(\operatorname{ad}(h_\mu)\right)(a)$, $\mu = \mu_1 \cup \mu_2 \cup \dots \cup \mu_s$.

In both cases lemma 4 applies and lemma 5 follows.

Now we are in a position to prove theorem 5.

Proof. Let $\mathfrak{B}$ be a subvariety of $\mathfrak{A}^3 \cap \mathfrak{N}_p\mathfrak{A}$ determined as a subvariety by a system $\{u_s \equiv 0, s = 1, 2, \dots\}$ of identical relations, where

$$u_s = \left(\operatorname{ad}([x_1, x_2])\right)^{p-1} ([x_{s+2}, x_{s+1}, \dots, x_2, x_1]) \,. \tag{15}$$

We show that $\mathfrak{B}$ is not finitely based. For, were this false, there existed an n, such that $u_n \equiv 0$ would follow *within* $\mathfrak{A}^3 \cap \mathfrak{N}_p\mathfrak{A}$ from a system $\{u_s \equiv 0, 1 \leqq s \leqq n-1\}$. But if $C_n \subseteq B_n \in \mathfrak{A}^3 \cap \mathfrak{N}_p\mathfrak{A}$ is an algebra introduced before lemma 5, then $u_s \equiv 0$, $0 \leqq s \leqq n-1$, hold in C_n, but $u_n \equiv 0$ does not.

Indeed, we prove, first, that $u_n(C_n) \neq \{0\}$:

$$u_n(a_\emptyset^{(1)}, h_\emptyset, g_{\{1\}}, h_{\{2,3\}}, h_{\{4,5\}}, \dots, h_{\{2n-2,2n-1\}}) = \pm \left(\operatorname{ad}(a_\emptyset^{(1)})\right)^{p-1} a_{T_n}^{(1)} =$$
$$= \pm b \neq 0 \,.$$

We prove, second, that $u_s(C_n) = \{0\}$, $0 \leqq s \leqq n-1$. Using lemma 5 it suffices to show that

$$\left(\operatorname{ad}(a_\sigma^{(1)})\right)^{p-1} ([c_1, c_2, \dots, c_s, a_\sigma^{(1)}]) = 0 \,, \quad c_i \in T_n((13)) \,, \quad i = 1, 2, \dots, s \,. \tag{16}$$

This is evident if $\sigma \neq \emptyset$. If $\sigma = \emptyset$ then either (16) equals zero or has the form $a_\tau^{(p)}$, where $|\tau| \leqq 2s \leqq 2n-2 < 2n-1$ by the choice of T. But by the definition of A_n, $a_\tau^{(p)} = 0$ if $|\tau| < 2n-1$.

Thus the proof of the theorem is complete.

Corollary. *Over any field k of prime characteristic p there exist uncountably many varieties of Lie algebras.*

Proof. Denote by $L = L(X, \mathfrak{A}^3 \cap \mathfrak{N}_p\mathfrak{A})$ the free Lie algebra of $\mathfrak{U} = \mathfrak{A}^3 \cap \mathfrak{N}_p\mathfrak{A}$ with a countable set X of free generators over k. Since $\mathfrak{U}$ is determined by the multilinear identities

$$[[[x_1, x_2], [x_3, x_4]], [[x_5, x_6], [x_7, x_8]]] \equiv 0 ,$$
$$[[x_1, x_2], [x_3, x_4], \dots , [x_{2p+1}, x_{2p+2}]] \equiv 0 ,$$

L is graded by the degrees of its elements in X.

Let now $\mathfrak{S}$ be the set of all subsets of positive integers. Consider for any $S \in \mathfrak{S}$ a subvariety $\mathfrak{W}_S$ of $\mathfrak{U}$ determined by the sets $\{u_s \equiv 0,\ s \in S\}$ of identical relations (see (15)). Let further W_S be the verbal subalgebra of L corresponding to $\mathfrak{W}_S$. To prove the corollary it suffices to show that $S \neq T$ implies $W_S \neq W_T$. Indeed, let $s \in (S \setminus T) \cup (T \setminus S) \neq \emptyset$, then since L is graded and u_s is multihomogeneous, if u_s is implied by some set $\{u_q,\ s \neq q \in Q\}$, then u_s is implied by $\{u_q,\ q < s,\ q \in Q\}$. This, however, is impossible by the final part of the proof of theorem 5. Hence the corollary is proved.

3.7. Wreath products

Let $\mathfrak{B}$ be a variety of Lie algebras over some ring k, A be an algebra from $\mathfrak{B}$, B be some algebra. A Lie algebra W is called a *wreath $\mathfrak{B}$-product* of the algebras A and B (denoted by $W = A\operatorname{wr}_{\mathfrak{B}}B$) if the following three conditions hold:

(i) $W = \operatorname{alg}(A, B)$.

(ii) Let $K = \operatorname{id}_W(A)$ be the ideal of W generated by A. Then $K \in \mathfrak{B}$.

(iii) If $\varphi: A \to C$, $\psi: B \to C$ are two homomorphisms of A, B into a Lie algebra C over k such that

(a) $C = \operatorname{alg}(\varphi(A), \psi(B))$;

(b) the ideal $S = \operatorname{id}_C(\varphi(A))$ generated by $\varphi(A)$ in C belongs to $\mathfrak{B}$,

then there exists a homomorphism $\chi: W \to C$ such that $\chi|_A = \varphi$, $\chi|_B = \psi$.

The existence and the most important properties of wreath $\mathfrak{B}$-products in the case of zero characteristic fields may be obtained as follows. Let $P = A * B$ be the free product of A by B (see 2, 6), Q be its ideal generated by A, $V(Q)$ be the verbal subalgebra of Q determined by the identities of $\mathfrak{B}$. By theorem 4.3, $V(Q)$ is an ideal of P and we may form the quotient algebra $P/V(Q)$. Then $W \cong P/V(Q)$, for (i) and (ii) follow by the construction and (iii) is implied by the universal property of the free product (see proposition 2, 6.1) and the definition of a verbal subalgebra.

Proposition 1. *Let $W = A\operatorname{wr}_{\mathfrak{B}}B$, then W is a split extension of B by $K = \operatorname{id}_W(A)$. Moreover if A is $\mathfrak{B}$-free then so is K.*

Proof. All statements follow from propositions 2, 6.1—2.

By the repeated use of proposition 2, 6.2 one may obtain a system of $\mathfrak{B}$-free generators of K, if one chooses a totally ordered basis E of B over k, and a system X of $\mathfrak{B}$-free generators of A. Then

$$[e_1, e_2, \dots , e_s, x],\ e_1 \geqq e_2 \geqq \dots \geqq e_s,\ e_i \in E,\ x \in X,\ s \geqq 0, \tag{1}$$

is a system of $\mathfrak{B}$-free generators of K.

Examples. (i) Let A, B be two algebras with bases X and E, E being totally ordered, A being abelian, and $W = A\operatorname{wr}_{\mathfrak{A}}B$ (or simply $A\operatorname{wr}B$ which is common if $\mathfrak{B} = \mathfrak{A}$, the variety of abelian algebras). Then K is an abelian algebra with a basis of the form (1). Consider K as a B-module. Then by the final remark from 2, 4, K is a U-module, $U = U(B)$, the universal envelope of B. A basis of $U(B)$ being formed by ordered E-monomials $e_1 e_2 \dots e_s$, $e_i \in E$, $s \geqq 0$, one immediately sees that K is a free U-module freely generated by X.

(ii) More specifically, let B be itself abelian. Then $U = U(B)$ is a polynomial algebra $k[E]$, and thus K is a free module over a polynomial algebra $k[E]$ with a system X of free generators. This situation admits generalization to any field k. Namely, one may prove the following instrumental

Theorem 6. *Let k be a ring and $M = L(X, \mathfrak{A}^2)$ be a free metabelian algebra over k. Assume that $x \mapsto y \mapsto z$ gives a bijection of the sets X, Y, Z. Let further K be a free $k[Z]$-module freely generated by Y, and $W = K \curlywedge B$ be a semidirect product of a free abelian algebra B freely generated by Z and K. Let also φ be a homomorphism of M into W extending the mapping $x \mapsto y + z$. Then φ is injective; moreover an element $w = b + \sum_{x \in X} f_x x$ belongs to the image of φ if and only if $b = \sum_{x \in X} f_x z$ ($f_x \in k[Z]$).*

Proof. $L(X)$ being a free Lie algebra over k, a system of free generators of $L(X)^2$ was found in lemma 2, 3.2 (ii) in the form $[x_1, x_2, \dots, x_{s-1}, x_s]$, $x_1 \geqq x_2 \geqq \dots \geqq x_{s-1} < x_s$, $x_i \in X$, $s \geqq 2$, X being in some way totally ordered. Clearly then the basis of $M = L/(L^2)^2$ is formed by the same elements but without restriction on s. We transfer the ordering of X onto Y and Z by the above given bijections and calculate the image of the basic monomial of M in W.

$$\varphi([x_1, x_2, \dots, x_{s-1}, x_s]) = [y_1 + z_1, y_2 + z_2, \dots, y_{s-1} + z_{s-1}, y_s + z_s] =$$
$$= [z_1, z_2, \dots, z_{s-2}, y_{s-1}, z_s] + [z_1, z_2, \dots, z_{s-2}, z_{s-1}, y_s] =$$
$$= z_1 z_2 \dots z_{s-2} z_{s-1} y_s - z_1 z_2 \dots z_{s-2} z_s y_{s-1} \text{ if } s \geqq 2, \text{ and } y_1 + z_1 \text{ if } s = 1 .$$

(The calculation uses the fact that K is an abelian ideal of W.) Considering the lexicographical ordering $\leqq$ on ordered X, Y and Z-monomials, we see that in our case $z_1 z_2 \dots z_{s-2} z_{s-1} \lneqq z_1 z_2 \dots z_{s-2} z_s$ for $z_s > z_{s-1}$. Therefore a nontrivial linear dependence of $\varphi(w_1), \dots, \varphi(w_m)$, w_i being some basic monomials in M, yield that of 'junior' terms of $\varphi(w_1), \dots, \varphi(w_n)$. However these junior terms determine w_i uniquely which gives a nontrivial linear dependence of w_i themselves. The details being left to the reader, we have proved so far the injectivity of φ.

We prove now the criterion. Note first that if w' is another element of the form $w' = b' + \sum f'_x x$, and $b = \sum f_x z$, $b' = \sum f'_x z$, then the same relation holds for $w + w'$ and $[w, w']$ by the immediate calculation: $[w, w'] = 0 + \sum_{x, x'} (f'_{x'} z' f_x - f_{x'} z' f'_x)\, x$. Clearly then

$$\sum_{x, x'} (f'_{x'} z' f_x - f_{x'} z' f'_x)\, z = \sum_{z, z'} (f'_{z'} z' f_z z - f_{z'} z' f'_z z) = 0 .$$

Since however $\varphi(x) = y + z$, $x \in X$, satisfies this relation, the whole of $\varphi(M)$ does the same. It remains to prove therefore that, given an element $w = b + \sum f_x x$ such that $b = \sum_x f_x z$, then $w \in \varphi(M)$. Clearly only the case $b = 0$ may be considered (for, otherwise, a suitable element from $\varphi(M)$ may be subtracted). Thus let $w = f_1 x_1 + f_2 x_2 + \dots + f_n x_n$, where all x_i are different, f_i are nonzero and n is minimal for such representations of elements with $\pi(w) = f_1 z_1 + f_2 z_2 + \dots + f_n z_n = 0$ and outside of $\varphi(M) \cap K$. If $f_i = f'_i z_n + f''_i$, then $f''_1 z_1 + f''_2 z_2 + \dots f''_{n-1} z_{n-1} = 0$. By our assumption then $w'' = f_1''x_1 + f_2''x_2 + \dots + f_{n-1}''x_{n-1} \in \varphi(M) \cap K$ and hence all $f_1, f_2, \dots, f_{n-1}$ may be considered in the form $f_1'z_n, f_2'z_n, \dots, f_{n-1}'z_n$. But an element $u = f_1'z_nx_1 - f_1'z_1x_n \in \in \varphi(M) \cap K$ and thus $\pi(w - u) = 0$. On the other hand $w - u = f_2x_2 + \dots + + (f_n + f_1'z_1)\, x_n$, which either contradicts the minimality of n or implies $w - u \in \in \varphi(M) \cap K$, hence $w \in \varphi(M) \cap K$.

This completes the proof of the whole theorem.

Interesting applications of this theorem will be shown in the next section, we prove next a result on the lower central series of the wreath product which will be used later.

Proposition 2. *Let B be a nilpotent algebra, A be a $\mathfrak{V}$-free algebra, $\mathfrak{V}$ being a variety of Lie algebras over a zero characteristic field k, $W = A\mathrm{wr}_{\mathfrak{V}}B$ be their wreath $\mathfrak{V}$-product. Then $\bigcap_{n=1}^{\infty} W^n = \{0\}$.*

Proof. Let c be the nilpotency class of B and choose a basis E of B in the form $E = E_1 \cup E_2 \cup \dots \cup E_c$, where E_i is a basis of B^i modulo B^{i+1}, $i = 1, 2, \dots, c$. Order E totally arbitrarily subject to a single condition: if $e \in E_i$ and $f \in E_j$, $i < j$, then $e < f$. By the PBW-theorem 2,4.6 a basis of $U(B)$ is then formed by the ordered monomials $a = e_1e_2 \dots e_s$, $e_1 \geqq e_2 \geqq \dots \geqq e_s$, $s \geqq 0$. Define also the weight $w(x)$ of an element $x \in U(B)$ by the rule: if $e \in E_i$ then $w(e) = i$, $w(a) = w(e_1) + \dots + w(e_s)$, $w(x)$ is the minimum of the weights of the nonzero basic summands of x. Let $U^{(s)}$ be the set of all elements of U having a weight $\geqq s$, and zero. Then each $U^{(s)}$ is an ideal of U (for a thourough verification of this fact see proposition 4, 1.1) and $\bigcap_{s \geqq 1} U^{(s)} = \{0\}$.

Let next K be the ideal generated by A in W, which is the $\mathfrak{V}$-free Lie algebra freely generated by the elements (1). Then as is seen easily K/K^2 is a free left $U = U(B)$-module, the generators having the form $v = e_1e_2 \dots e_sx$. We define the weight of this free generator v by $w(v) = w(e_1e_2 \dots e_s) + 1$. Moreover, given a set R of basic monomials in this generating system, define inductively the weight $w(z) = w(z_1) + w(z_2)$ if $z = = [z_1, z_2] \in R$. Let finally K_s be the span of all basic monomials of K having weight $\geqq s$. Evidently, K_s is an ideal of K. Moreover, for each $e \in E$, $\mathrm{ad}(e)$ is a derivation of K and thus to prove that K_s is closed under $\mathrm{ad}(e)$, it suffices to prove that any free generator $e_1e_2 \dots e_tx$, $w(e_1e_2 \dots e_t) = r$ is mapped by $\mathrm{ad}(e)$ into K_{r+1}. This however is implied by the fact that $U^{(r)}$ is an ideal of U.

The reader is assumed to check up that W/K_s is a nilpotent algebra (use for example criterion 1, 10.6 where $N = K$). This means that each K_s contains some W^{n_s}. Since clearly $\bigcap_{s \geqq 1} K_s = \{0\}$ the same is true for $\bigcap_{s \geqq 1} W^s \supseteq \bigcap_{s \geqq 1} W^{n_s} = \{0\}$. Thus the proposition is proved.

3.8. Applications to solvable varieties

The aim of this section is twofold: first, to study derivations and automorphisms of free solvable algebras, and, second, to prove that the identical relations of any metabelian variety are finitely based.

Given a nilpotent derivation D (that is, $D^{s+1} = 0$) of a Lie algebra A over some ring k, where $s!$ is invertible, one may consider a linear mapping $\mathrm{e}^D = \sum_{t=0}^{s} \frac{D^t}{t!}$ which is in fact an automorphism of A. For

$$\begin{aligned}
e^D([x, y]) &= \left(\sum_{t=0}^{s} \frac{D^t}{t!}\right)([x, y]) = \sum_{t=0}^{s} \frac{1}{t!} \left(D^t([x, y])\right) = \\
&= \sum_{t=0}^{s} \frac{1}{t!} \sum_{u+v=t} \binom{t}{u} [D^u(x), D^v(y)] = \sum_{t=0}^{s} \sum_{u+v=t} \left[\frac{D^u}{u!}(x), \frac{D^v}{v!}(y)\right] = \\
&= \left[\sum_{u=0}^{s} \frac{D^u}{u!}(x), \sum_{v=0}^{s} \frac{D^v}{v!}(y)\right] = [e^D(x), e^D(y)] .
\end{aligned}$$

If $D = \mathrm{ad}(a)$ for some $a \in A$, then $e^D = e^{\mathrm{ad}(a)}$ is called an *inner automorphism* induced by the element a.

Recall also that if $\mathfrak{B} = \mathfrak{A}^l$ is the variety of all solvable Lie algebras of solvable length $\leq l$, then its free algebra $L(X, \mathfrak{B}) = G$ is called a *free solvable algebra* of solvable length l. Clearly then $G^{(l-1)} \neq \{0\}$, $G^{(l)} = \{0\}$.

Theorem 7 (Šmel'kin). *Let $G = L(X, \mathfrak{A}^l)$, $|X| > 1$, be a free solvable algebra of solvable length l, φ its automorphism such that $\varphi|_{G^{(l-1)}} = 1_{G^{(l-1)}}$, D its derivation with $D|_{G^{(l-1)}} = 0_{G^{(l-1)}}$. Then there exist $a, b \in G^{(l-1)}$ such that $\varphi = e^{\mathrm{ad}(a)}$, $D = \mathrm{ad}(b)$.*

Proof. Consider first the case $l = 2$, that is the case of a free metabelian algebra G. Using theorem 6.6 we denote by the same letter a subalgebra of $W = K \curlywedge B$, where B is a free abelian algebra freely generated by Z, K is a free $k[Z] = U(B)$-module freely generated by Y. Thus G is generated by $y + z$, $y \in Y$, $z \in Z$. Clearly, $G^2 = G \cap K$.

Now we make two simple remarks. First, note that $C = C_G(G^2) = G^2$. For otherwise $\alpha_1 x_1 + \alpha_2 x_2 + \dots + \alpha_n x_n \in C$ with, say, $\alpha_1 \neq 0$. Let then S be a two-dimensional metabelian algebra spanned by f, e with $[f, e] = e$ (see example 1, 5 (ii)). We map then $x_1 \mapsto f$, $x_2 \mapsto e$, $x \mapsto 0$, $x \neq x_1, x_2$. By our assumption then $\alpha_1 f + \alpha_2 e$ centralizes $[f, e] = e$ which is impossible. Second, note that given an ideal N such that $\varphi|_N = 1_N$, $D|_N = 0_N$ and $C = C_G(N)$, then $\varphi|_{G/C} = 1_{G/C}$ and $D|_{G/C} = 0_{G/C}$. For, given $g \in G$, $x \in N$, one has $[g, x] = \varphi([g, x]) = [\varphi(g), x]$ and $0 = D([g, x]) = [D(g), x] + [g, D(x)] = [D(g), x]$. In both cases $\varphi(g) - g$ and $D(g)$ are elements of C which proves our remark.

Consider first the case of φ. By our remarks φ is identical on G/G^2 and therefore $\varphi(x) = x + u_x$, $u_x \in G^2$. Thus one has $[x, x'] = \varphi([x, x']) = [x + u_x, x' + u_{x'}] = [x, x'] + [x, u_{x'}] - [x', u_x]$, whence $[x, u_{x'}] = [x', u_x]$, for any $x, x' \in X$, or in K one has $u_{x'} = z't$, $t \in K$ and $zz' \cdot t = z' \cdot u_x$. Now $u_x = z \cdot t = [x, t]$ for all $x \in X$. By the criterion of theorem 6.6, $t \in G$. Therefore $\varphi(x) = x + [x, t] = e^{\mathrm{ad}(t)}(x)$ for all $x \in X$ (clearly $\mathrm{ad}(t)^2 = 0$). This proves $\varphi = e^{\mathrm{ad}(t)}$, $t \in G^2$. Quite similarly $D = \mathrm{ad}(b)$, $b \in G^2$.

Now assume $l > 2$. Then by theorem 2,3.3, $G^{(l-2)} = L^{(l-2)}/L^{(l)}$, L being a free Lie algebra freely generated by X, is a free metabelian algebra. By the just proved case $l = 2$, $\varphi|_{G^{(l-2)}} = e^{\mathrm{ad}(a)}$, $a \in G^{(l-1)}$, so that $\varphi\psi^{-1}$ is identical on $G^{(l-2)}$. By the remark above then $\varphi\psi^{-1}$ is identical on $G/C_G(G^{(l-2)})$. However, as easily seen, $C_G(G^{(l-2)}) = \{0\}$ and thus $\varphi = \psi$ is an inner automorphism. The same argument being valid for derivations, the proof is complete.

In the remaining part of this section we sketch the proof of theorem 8 whose group theoretical version is due to D. E. Cohen.

Theorem 8. *Let $\mathfrak{B}$ be a metabelian variety of Lie algebras over some noetherian ring k. Then the identical relations holding in $\mathfrak{B}$ are implied by a finite subset of them. In other words, $\mathfrak{B}$ is finitely based.*

In fact some stronger result may be proved. To formulate it, remark that any metabelian variety $\mathfrak{B}$ is uniquely determined by a verbal subalgebra $V(L)/L''$ of a free metabelian Lie algebra $M(X) = L(X)/(L(X))''$ of countable rank, $V(L)$ being the verbal subalgebra of $L = L(X)$ determining $\mathfrak{B}$, L'' being the second derived algebra of L. Let X be formed by the elements $x_1, x_2, \dots, x_n, \dots$, N be the set of all positive integers and Φ be the set of all one-to-one order preserving mappings of N into itself. An ideal $B \subseteq M$ is called a *Φ-subalgebra* if B is invariant under all endomorphisms of $M(X)$ induced by any mapping from Φ. Any verbal subalgebra being obviously a Φ-subalgebra, it suffices to prove

Proposition 1. *Any Φ-ideal B of a free metabelian algebra $M = M(X)$ of countable rank is finitely generated as a Φ-ideal.*

Reducing the proof of this proposition entirely to commutative algebra and ordered set theory is the following argument. Let S be a polynomial ring $S = k[z_1, z_2, \dots, z_n, \dots]$ and K be an S-module freely generated by the set $Y = \{y_1, y_2, \dots, y_n, \dots\}$. Any element a from some *ideal* I of S (respectively *submodule* T of K) having the form $f(z_1, z_2, \dots, z_n)$ (respectively $\sum_i f_i(z_1, \dots, z_n)\, y_i$), an ideal I (respectively submodule T) is referred to as a *Φ-ideal* (respectively *Φ-submodule*) if, for any $\varphi \in \Phi$, one has $f(z_{\varphi(1)}, z_{\varphi(2)}, \dots, z_{\varphi(n)}) \in I$ (respectively, $\sum_i f_i(z_{\varphi(1)}, \dots, z_{\varphi(n)})\, y_{\varphi(i)} \in T$). Using theorem 6.6 imbed M into a wreath product $W = A \operatorname{wr} \widetilde{B}$, where $A = A(Y)$, $\widetilde{B} = \widetilde{B}(Z)$ be free abelian algebras generated by Y, Z. By the construction, W is a split extension of B by K, the latter ideal being a free $U(B)$ $(= S)$-module with basis Y, and $M = \operatorname{alg}\{y_i + z_i, i \in \mathsf{N}\}$. Consider, without loss of generality, a Φ-subalgebra B *contained in* $M^2 \subseteq K$. Taking into account that the projection of M onto $\widetilde{B}$ equals this latter, B is an S-submodule of K, and a straightforward verification shows that B is finitely generated as a Φ-submodule if and only if B is finitely generated as a Φ-subalgebra, thus supplying the reduction of the problem to commutative algebra.

Denote now by V the set of all natural valued sequences $(i_1, i_2, \dots, i_n)$ with finite support. One may endow V by the total lexicographical ordering $\leqq$ on one hand, and by the partial ordering $\lesssim$ defined through $(i_1, i_2, \dots, i_n) \lesssim (j_1, j_2, \dots, j_m) \Leftrightarrow n \leqq m$ and for some $\varphi \in \Phi$, $i_r \leqq j_{\varphi(r)}$, where $1 \leqq r \leqq n$ and $1 \leqq \varphi(r) \leqq m$, on the other. Again the set $V \times \mathsf{N}$ may be ordered in two ways, first, by putting $((i_1, \dots, i_n), r) \leqq ((j_1, \dots, j_m), s) \Leftrightarrow$ $\Leftrightarrow r < s$ or $r = s$ and $(i_1, \dots, i_n) \leqq (j_1, \dots, j_m)$, second by putting $((i_1, \dots, i_n), r) \lesssim ((j_1, \dots, j_m), s) \Leftrightarrow$ for some $\varphi \in \Phi$, $\varphi(r) = s$, $\varphi(n) \leqq m$, $(i_1, \dots, i_n) \lesssim (j_1, \dots, j_m)$. Finally consider the set $Q = k \times V \times \mathsf{N}$ and call its subset $\operatorname{cl}(X)$ a *closure* of the subset $X \subseteq Q$ if $\operatorname{cl}(X)$ is a set of triples (c, v, n) such that for some $(c_1, v_1, n_1), \dots, (c_m, v_m, n_m) \in X$ one has $(v_i, n_i) \lesssim (v, n)$, $i = 1, 2, \dots, m$, and denoting by C the ideal generated by $c_1, c_2, \dots, c_m$, one has $c \in C$. Q is said to have the *finite basis property* if any of its closed subsets X $(= \operatorname{cl}(X))$ is the closure of some finite $Y \subseteq X$.

The second reduction promised will be obtained by corresponding a closed subset $\pi(B) \subseteq Q$ to each Φ-submodule B of K. To this end denote by $wt(f)$, where $f \in S$ and has the form $f = \sum a x_1^{i_1} \dots x_n^{i_n}$, an element $(a, (i_1, \dots, i_n)) \in k \times V$ whose projection onto V is lexicographically the largest among the monomials involved in f. Given now a nonzero element $a = \sum_{i=1}^{n} f_i y_i \in K$ with $f_n \neq 0$, put $\pi(a) = (wt(f_n), n)$. The reader being assumed to prove that $\pi(B)$ is a closed subset of Q, we prove that $\pi(B)$ possessing a finite basis implies B being finitely generated as a Φ-submodule of K. Thus assume $\pi(B) = \mathrm{cl}\{\pi(a_1), \pi(a_2), \dots, \pi(a_r)\}$ and choose an arbitrary $a \in B$. Then in the Φ-submodule of B generated by $a_1, a_2, \dots, a_r$ one may find an element b with $\pi(b) = \pi(a)$. Clearly, if $\pi(a) = (c, v, n)$, $\pi(a - b) = (c', v', n')$ with $(v', n') < (v, n)$. Since $\leqq$ is a full ordering of $V \times \mathsf{N}$, the proof of the statement is completed by induction.

Now we are in a position to overcome our last obstacle.

Proposition 2. *Q possesses the finite basis property.*

We anticipate the proof by one more definition. A partially ordered set $(P, \leqq)$ is called a *partially well ordered* set if any of its nonempty subsets, say X, contains a finite subset Y such that for any $s \in X$ there exists a $y \in Y$ with $y \leqq x$. Alternatively, P is partially well ordered if any sequence of its elements contains a (possibly non-strictly) increasing subsequence.

Lemma. *$(V \times \mathsf{N}, \leqq)$ is a partially well ordered set.*

$(V, \leqq)$ being partially well ordered is a theorem due to G. Higman. We omit its intuitively clear but far from trivial proof and show how this theorem implies our lemma. Let then $\{(\alpha_s, r_s)\}$ be an infinite sequence of elements in $V \times \mathsf{N}$. Taking its infinite subsequence with $r_1 \leqq r_2 \leqq \dots$, one may without loss of generality think this chain of inequalities holding for the original sequence. Given any element $(\alpha, r) \in V \times \mathsf{N}$ of the form $((i_1, i_2, \dots, i_{r-1}, i_r, i_{r+1}, \dots, i_m), r)$, say that $(i_1, \dots, i_{r-1})$ is an *initial segment*, $(i_{r+1}, \dots, i_m)$ a *terminal segment* and i_r a *pivotal element* of (α, r). To find a nondecreasing subsequence of (α, r) choose from the original sequence an infinite subsequence $\{(\alpha_s', r_s')\}$ of elements whose initial segments increase; then choose from $\{(\alpha_s', r_s')\}$ an infinite sequence $\{(\alpha_s'', r_s'')\}$ of elements whose pivotal elements increase, and finally an infinite subsequence of elements from $\{(\alpha_s'', r_s'')\}$ in which the terminal segments increase. The resulting sequence is one increasing in $(V \times \mathsf{N}, \leqq)$ and the lemma is proved.

We denote, finally, $V \times \mathsf{N}$ by P and show that $Q = k \times P$ possesses the finite basis property. Indeed, given a closed subset $X \subset Q$, consider for each $p \in P$ an ideal $X(p) \subset k$ equal to the projection of $X \cap (k \times p)$ onto k. X being closed yields $X(p) \subseteq X(q)$ provided that $p \leqq q$. Order $\{X(p), p \in P\}$ by $X(p) < X(q)$ if and only if $p \leqq q$ and $X(p) = X(q)$. Then $(\{X(p), p \in P\}, <)$ is easily seen to be partially well ordered and hence one may choose minimal elements $X(p_1), \dots, X(p_r)$ such that for any $p \in P$ there exist $i = 1, \dots, r$ with $X(p_i) < X(p)$. As k has the finite basis property each $X(p_i)$ is generated by a finite subset $\{c_{i1}, \dots, c_{it_i}\}$. Now it remains to verify $X = \mathrm{cl}\left\{\bigcup_j (c_{ij}, p_i), i = 1, \dots, r\right\}$ and the proof of the theorem is complete.

CHAPTER 4

Universal enveloping algebra Further results and applications

4.1. PI-algebras

An associative algebra R over some ring k is called a PI-*algebra* if R satisfies a nontrivial identity $f(x_1, x_2, \dots, x_d) \equiv 0$, where $f = f(x_1, x_2, \dots, x_d)$ is an element of a free associative algebra $A^+(X)$, $X \ni x_1, x_2, \dots, x_d$, over the same ring k. Identical relations in associative algebras are termed *polynomial* ones, which explains the notation of these latters.

We will not intervene deeply into the theory of PI-algebras full of brilliant theorems and difficult problems. The main question we shall be interested in is: what conditions are necessary and sufficient to provide that the universal envelope $U(G)$ of a Lie algebra G were a PI-algebra. The importance of this problem for Lie algebras themselves will be seen later, we start with a simple

Remark. Let G be a Lie algebra over some field k, $U(G)$ be its universal envelope. Assume that $U(G)$ is a PI-algebra and that $[x_1, x_2, \dots, x_{c+1}] \equiv 0$ is its identity. Then G is abelian.

We show, first, that $U(G)$ has no zero divisors. We form as in 2,4 the graded algebra $\operatorname{gr}(U(G))$ associated with $U(G)$. By the corollary to the PBW-theorem 2,4.6, gr $U(G)$ is isomorphic to a polynomial ring and thus has no zero divisors. Given now two nonzero elements $u, v \in U(G)$, let $\bar{u}, \bar{v} \neq 0$ be their images in $\operatorname{gr}(U(G))$. Then $uv = 0$ implies $\overline{uv} = \bar{u}\bar{v} = 0$ which is impossible.

It suffices now to prove that no noncommutative associative algebra R without zero divisors can be nilpotent as a Lie algebra, or to prove that under the same conditions $[x_1, x_2, \dots, x_{c+1}] \equiv 0$, $c \geqq 2$, implies $[x_1, x_2, \dots, x_c] = 0$ in R. So let $r_1, r_2, \dots, r_c$ be such that $r = [r_1, r_2, \dots, r_c] \neq 0$ and put $s = [r_2, \dots, r_c]$. Then for any $a \in R$, $[a, s]$ is central in R, whence

$$0 = [r_1, [sr_1, s]] = [r_1, s[r_1, s]] = [r_1, s]^2 = r^2 ,$$

contradicting the absence of zero divisors in R.

Before we go on to the main part of the chapter we prove in preliminaries some results relevant to another problem relying Lie and associative algebras with identities. Namely, call a Lie algebra G an SPI-*algebra* if G can be isomorphically embedded in some (associative) PI-algebra. Many Lie algebras are known to be SPI-algebras, it is as well known that not every Lie algebra with identity is an SPI-algebra so that the description of SPI-algebras is not complete. We prove below that any nilpotent Lie algebra over some field and any finite-dimensional Lie algebra over some prime characteristic field are SPI-algebras.

An associative algebra A is called *nilpotent of index* n if $x_1x_2 \dots x_n \equiv 0$ is an identity in A. Note that A is a nilpotent Lie algebra under bracketing of the nilpotency class $\leqq n - 1$.

Proposition 1. *Every nilpotent Lie algebra G over some field k can be embedded in a nilpotent associative algebra.*

Proof. As in 3.6.proposition 2, choose a basis E of G as follows. Given $(G^n)_{n \geqq 1}$, the lower central series of G, $G^c \neq \{0\}$, $G^{c+1} = \{0\}$, take for E_i, $i = 1, 2, \dots, c$, a set of elements of G, forming a basis of G_i modulo G_{i+1}. Then $E = \bigcup_{i=1}^{c} E_i$ is a basis of G. Order E totally arbitrarily subject to a single condition: if $e \in E_i$, $f \in E_j$, $i < j$, then $e < f$. By the PBW-theorem 2,4.6 the basis of $U^+(G)$ is formed by the ordered monomials of the form

$$a = e_1e_2 \dots e_s\,, \qquad e_1 \geqq e_2 \geqq \dots \geqq e_s\,, \qquad s \geqq 1\,. \tag{1}$$

We define the weight $w(x)$ of an element $x = \sum_{\alpha_i \neq 0} \alpha_i a_i \in U^+(G)$ by the rule: if $e \in E_i$, then $w(e_i) = i$, $w(a) = \sum_{i=1}^{s} w(e_i)$, $w(x)$ is the minimum of the weights $w(a_i)$ of a_i.

Let then $U^{(s)}$ be a set of elements $x \in U^+(G)$ such that $w(x) \geqq s$, $U^{(s)}$ being evidently a linear subspace of $U^+(G)$. To prove that $U^{(s)}$ is an ideal of U, it suffices to show that if $e_1, e_2, \dots, e_l \in E$ then $e_1e_2 \dots e_l \in U^{(w(e_1)+\dots+w(e_l))}$. We proceed by induction on l, the case $l = 1$ being trivial. Adapting the argument from the proof of the lemma 2,2.2, one notes that if $e_{i_1} \geqq e_{i_2} \geqq \dots \geqq e_{i_l}$, $\{i_1, i_2, \dots, i_l\} = \{1, 2, \dots, l\}$, then $e_1e_2 \dots e_l$ is representable as a sum of $e_{i_1}e_{i_2} \dots e_{i_l}$, having by the definition weight $w(e_{i_1}) + \dots + w(e_{i_l})$ and monomials of the form $e_{j_1} \dots [e_{j_r}, e_{j_{r+1}}] \dots e_{j_l}$ whose weight by the induction hypothesis $\geqq w(e_{j_1}) + \dots + w([e_{j_r}, e_{j_{r+1}}]) + \dots + w(e_{j_l}) \geqq w(e_1) + \dots + w(e_l)$ since by the choice of E and the inequality $[G^u, G^v] \subseteq G^{u+v}$ (proposition 1,7.2), $w([e_{j_r}, e_{j_{r+1}}]) \geqq w(e_{j_r}) + w(e_{j_{r+1}})$. Thus $U^{(s)}$ is an ideal of $U^+(G)$.

In particular $U^{(c+1)}$ is an ideal of $U^+(G)$ such that $G \cap U^{(c+1)} = \{0\}$. Denote by A the quotient algebra of $U^+(G)$ by $U^{(c+1)}$ and by ε the natural epimorphism $\varepsilon\colon U^+(G) \to A$. Since $\operatorname{Ker}(\varepsilon) \cap G = \{0\}$, $\varepsilon|_G$ is an embedding of G into A and a trivial observation of the nilpotency of A completes the proof of the proposition.

Proposition 2 (Iwasawa). *Let k be a field of prime characteristic $p > 0$ and G be a finite dimensional Lie algebra over k. Then G is embeddable in a finite dimensional associative algebra over k.*

Lemma 1. *Let* $\dim G = n$. *Then $U(G)$ is a free module of finite type over its central subalgebra $R = k[y_1, y_2, \dots, y_n]$, the polynomial ring in n variables.*

Proof. Let $e_1, e_2, \dots, e_n$ form a basis of G over k; then, by the PBW-theorem 2,4.6 a basis of $U(G)$ may be formed by

$$e_1^{l_1}e_2^{l_2} \dots e_n^{l_n}\,, \qquad l_i = 0, 1, 2, \dots, \qquad i = 1, 2, \dots, n\,. \tag{2}$$

Consider then $\varphi_i = (\operatorname{ad}(e_i)) \in \operatorname{End}_k(G) = \Phi$; $\dim \Phi = n^2$, so that for any $r \geqq n^2$, the elements $\varphi_i, \varphi_i^p, \dots, \varphi_i^{p^r}$ must be linearly dependent, that is there exist elements $\alpha_0, \alpha_1, \dots, \alpha_r \in k$ not all zero with $\alpha_0\varphi_i + \alpha_1\varphi_i^p + \dots + \alpha_r\varphi_i^{p^r} = 0$. Denote by $f_i = f_i(t)$ a polynomial $f_i(t) = \alpha_0 t + \alpha_1 t^p + \dots + \alpha_r t^{p^r}$, thus $f_i(\operatorname{ad}(e_i)) = 0$, $i = 1, 2, \dots, n$. By the proposition 1,14.1(b), $(\mathrm{ad}(e_i))^p = \mathrm{ad}(e_i^p)$ and therefore

$$f_i(\mathrm{ad}(e_i)) = \sum_{j=0}^{r} \alpha_j(\mathrm{ad}(e_i))^{p^j} = \sum_{j=0}^{r} \alpha_j \,\mathrm{ad}(e_i^{p^j}) = \mathrm{ad}(f(e_i)) .$$

Given an element $x \in G$, $[f_i(e_i), x] = (\mathrm{ad}(f_i(e_i)))\,(x) = (f_i(\mathrm{ad}(e_i)))\,(x) = 0$; moreover, since G generates $U(G)$, $f_i(e_i)$ commutes with all elements of $U(G)$ and hence is central. We denote $f_i(e_i)$ by y_i, $i = 1, 2, \ldots, n$, and prove that if d_i is the degree of f_i, the basis of $U(G)$ may be also formed by monomials

$$e_1^{l_1}e_2^{l_2} \ldots e_n^{l_n}y_1^{s_1}y_2^{s_2} \ldots y_n^{s_n}, \qquad 0 \leqq l_1 < d_i , \qquad 0 \leqq s_i , \tag{3}$$

$i = 1, 2, \ldots, n$. For clearly (3) generate $U(G)$ as a k-module; moreover any finite subset of monomials of the form (3) lies in a subspace $V_{q_1q_2\ldots q_n}$ generated by all the monomials of the form (3), such that $s_i \leqq q_i$, $i = 1, 2, \ldots, n$. But the same subspace is spanned by monomials (2) with $l_i \leqq (q_i + 1)\, d_i - 1$, $i = 1, 2, \ldots, n$. An immediate calculation shows that the number of monomials of both systems in $V_{q_1q_2\ldots q_n}$ is the same; then the linear independence of (2) implies that of (3).

In particular a subsystem of (3) formed by the monomials of the form $y_1^{s_1}y_2^{s_2} \ldots y_n^{s_n}$, $s_i \geqq 0$, $i = 1, 2, \ldots, n$, spans a central subalgebra R evidently isomorphic to a polynomial ring $k[y_1, y_2, \ldots, y_n]$ such that $U(G)$ is a free R-module of rank $d_1d_2 \ldots d_n$. The proof is thus complete.

The proof of this important lemma forming the main part of the proof of the proposition, now we can complete this latter in a few lines.

Proof of the proposition. Let I be a subspace of $U(G)$ spanned by the monomials (3) for which some $s_i \neq 0$, $i = 1, 2, \ldots, n$. Clearly I is an ideal of $U(G)$, a direct summand of a k-module $U(G)$ with a direct complement E spanned by a finite number of those monomials (3) for which $s_1 = s_2 = \ldots = s_n = 0$. Since all d_i may be chosen $\geqq 2$, $G \subseteq E$ and thus the natural epimorphism $\varepsilon\colon U(G) \to U(G)/I$ maps G isomorphically, the latter associative algebra, generated by $\varepsilon(E)$, being finite dimensional. So we are finished.

To prove the last corollary in this section we need one more definition. Let n be a positive integer, P_n be the set of all substitutions of $\{1, 2, \ldots, n\}$, $|\sigma| = 1$ if the substitution $\sigma \in P_n$ is even and $|\sigma| = -1$ otherwise. Then the identical relation

$$S_n(x_1, x_2, \ldots, x_n) = \sum_{\sigma \in P_n} |\sigma|\; x_{\sigma(1)}x_{\sigma(2)} \ldots x_{\sigma(n)} \equiv 0 \tag{4}$$

is called a *standard identity* of the degree n. The reader easily checks up that if R is of finite type $< n$ over some ring k, then (4) holds in R identically.

Now we formulate the promised

Corollary. *Let G be a finite dimensional Lie algebra over a field k of prime characteristic p. Then G is an SPI-algebra.*

This immediately follows from proposition 2 and the remark on standard identities.

4.2. Modules and identities. Zero characteristic case

We prove first the auxiliary but very useful

Proposition 1. *Let k_n be the ring of n by n matrices over some commutative ring k with unity. Then k_n satisfies no polynomial identity of degree $< 2n$.*

Proof. By proposition 3, 3.1, reworded for the associative algebra case, given a nontrivial identity of degree d in k_n, there exists a multilinear identity of degree d valid in k_n. Thus assume without loss of generality that k_n satisfies an identity

$$x_1 x_2 \dots x_d \equiv \sum_{\sigma \neq 1} a_\sigma x_{\sigma(1)} x_{\sigma(2)} \dots x_{\sigma(d)} \tag{1}$$

with $d < 2n$. Denote by E_{sr} a matrix from k_n with entries e_{ij}, $1 \leqq i, j \leqq n$, such that $e_{sr} = 1$ and $e_{ij} = 0$ if $i \neq s$ or $j \neq r$. Consider the values of both sides of (1) in $x_1 = E_{11}$, $x_2 = E_{12}$, $x_3 = E_{22}$, ... Then, since $d < 2n$, the left hand side of (1) is nonzero, while the right one is always zero, the routine verification of both assertions being left to the reader.

Next we consider an

Example I. Let G be a Lie algebra over a field k of zero characteristic with basis x, y, z and multiplication table $[x, y] = z$, $[x, z] = [y, z] = 0$. Let further $M = k[t]$ be a polynomial ring over k. Then using example 1, 5 (i), M may be considered as a G-module through $xf = f'$, $yf = tf$, $zf = f$, f be a polynomial from M. As it was noted in the above mentioned example, M is an irreducible G-module of infinite dimension. Let A be the associative subalgebra of $\mathrm{End}_k(M)$ generated by x, y and $z = 1$. Our purpose is to show that for each positive n, A contains an associative subalgebra which can be mapped homomorphically onto k_n. So far, let V be the subspace of M spanned by $1, t, t^2, \dots, t^{n-1}$, B be the subalgebra of A consisting of all elements from A leaving V invariant, $\varphi: B \to \mathrm{End}_k(V)$ given by $\varphi(\mathcal{A}) = \mathcal{A}|_V$, $\mathcal{A} \in B$.

We have to show that φ is epi. To do this we construct a sequence of transformations $\mathcal{E}_{ij}^{(r)} \in \mathrm{End}_k(M) \cap A$, $r = 0, 1, 2, \dots$, such that $\mathcal{E}_{ij}^{(r)}(t^s) = 0$, $s < i$; t^j, $s = i$; 0, $i < s \leqq i + r$; $\alpha_{sr} t^{s-i+j}$, $s > i + r$. Put $\mathcal{E}_{ij}^{(0)} = \frac{1}{i!} (y^j x^i - y^{j+1} x^{i+1})$; the reader may verify that the above written conditions satisfy. Having constructed already $\mathcal{E}_{ij}^{(r)}$ and $\mathcal{E}_{ml}^{(0)}$ for all pairs m, l, put $\mathcal{E}_{ij}^{(r+1)} = \mathcal{E}_{ij}^{(r)} - \alpha_{i+r+1,r} \mathcal{E}_{i+r+1,j+r+1}^{(0)}$. Evidently, the conditions for $\mathcal{E}_{ij}^{(r+1)}$ are satisfied. Now the transformation $\mathcal{E}_{ij}^{(n-i)} \in B$ for all $1 \leqq i, j \leqq n$, and $\varphi(\mathcal{E}_{ij}^{(n-i)})$ is a linear transformation of V which maps $t^i \mapsto t^j$ and $t^s \mapsto 0$, $s \neq i$. Since such transformations span $\mathrm{End}_k(V)$, φ is surjective and our assertion follows.

Lemma 1. *Let G be a three dimensional nonabelian nilpotent algebra over a field k of characteristic zero, $U = U(G)$ be its universal envelope. Then U is not a PI-algebra.*

For, clearly G is isomorphic to the algebra of the preceding example; then by the definition of the universal envelope, the associative algebra A of this example is a homomorphic image of U. Having supposed that U is a PI-algebra, we should obtain that A and therefore all its homomorphic images satisfy a nontrivial identity of degree, say d, which is impossible by proposition 1, for among these homomorphic images one has k_d. Thus the proof is complete.

We show now that the choice of G in the example is in a sense not accidental.

Lemma 2. *Let G be a nonabelian finite dimensional algebra over some algebraically closed field of characteristic zero. Then either G contains a three dimensional nonabelian nilpotent subalgebra, or G contains a two dimensional nonabelian metabelian subalgebra.*

Proof. This is an easy exercise for, if G is not nilpotent, then for some $x \in G$ not all eigenvalues of $\mathrm{ad}(x)$ are zero; for example, let α be such a nonzero eigenvalue and y be a corresponding eigenvector. Then $H = \mathrm{alg}(\alpha^{-1}x, y)$ is a two dimensional metabelian nonabelian subalgebra. The case of nilpotent G being even more straightforward, the proof is complete.

The just proved lemma causes us to consider an

Example II. Let H be the Lie algebra spanned by u, v over a field k of characteristic zero subject to the multiplication rule $[u, v] = u$. Turn $M = k[t]$ into an H-module by putting $ut^r = (t + 1)^r$, $vt^r = t^{r+1}$, $r = 0, 1, 2, \ldots$. Again as in the previous example I, M is an irreducible H-module. Let C be an associative subalgebra of $\mathrm{End}_k(M)$ generated by $1, u, v$. Putting $x = uv - v$, $y = v$, $z = 1$, we can carry out the argument of example I without changes. This proves that C and, further, $U(H)$ can satisfy no nontrivial polynomial identity.

Now we can prove

Theorem 1. *Let G be a finite dimensional Lie algebra over some field k of characteristic zero. Then the universal envelope $U(G)$ of G is a PI-algebra if and only if G is abelian.*

Proof. In the light of the preceding results only the reduction to an algebraically closed field merits mention. However it is not a difficult task to check up that if k^* is an algebraic closure of k, then $U(G) \otimes_k k^*$ is a universal envelope of $G_{k^*} = G \otimes_k k^*$. Moreover, having assumed that $U(G)$ is a PI-algebra, there exists a nontrivial multilinear identity holding in $U(G)$. Direct calculations show that $U(G) \otimes_k k^*$ is a PI-algebra. Now the preceding lemmas and examples prove that G_{k^*} and hence G are abelian. Thus the theorem follows.

Assume now that G is infinite dimensional over k and that $U(G)$ is a PI-algebra. Then, since $U(G)$ is semisimple and by a theorem due to I. Kaplansky [56], $U(G)$ is a subdirect product of primitive algebras U_α having finite dimensions over their centres $k_\alpha \supset k$, k_α being a field. Let π_α be a projection of G onto U_α. Then $\pi_\alpha(G)\, k_\alpha$ is a finite dimensional Lie algebra over k_α. A homomorphism $\pi_\alpha \otimes 1_{k_\alpha}\colon G \otimes_k k_\alpha \to \pi_\alpha(G)\, k_\alpha$ induces an epimorphism of $U(G) \otimes k_\alpha$ onto the universal envelope $U_{k_\alpha}(\pi_\alpha(G)\, k_\alpha)$, which is therefore a PI-algebra. Since, further, $\pi_\alpha(G)\, k_\alpha$ is finite dimensional, $\pi_\alpha(G)$ is abelian, and U_α is commutative, α being arbitrary. Hence $U(G)$ is commutative and so we have proved

Theorem 1' (Latyšev). *Let k be a zero characteristic field, G be a Lie algebra over k, $U(G)$ be a PI-algebra. Then G is abelian.*

4.3. Modules and identities. Prime characteristic case

Contrary to the zero characteristic case, where finite dimensional algebra possesses as a general rule infinite dimensional simple modules, in the prime characteristic case one has

Theorem 2 (Curtis). *Let G be a finite dimensional Lie algebra over some field k of prime characteristic $p > 0$ and M be some irreducible G-module. Then M is finite dimensional, moreover, if k is algebraically closed, then the dimensions of irreducible G-modules are bounded.*

Proof. Remark first that if M is some G-module, then, by 2, 4, M is, say, left $U(G)$-module, moreover, given an irreducible G-module M, M is irreducible as a U-module.

All these assertions immediately follow from the universal property defining $U = U(G)$.

Recall that U is a free module of finite type over its central subalgebra $A = k[y_1, y_2, \dots, y_n]$, $n = \dim(G)$ (proposition 1.2). Since M is simple, it is cyclic and therefore isomorphic to a quotient module U/I of U by some maximal left ideal I. Let K be a subring of U formed by elements x such that $Ix \subseteq I$. Then I is a two-sided ideal of K.

Lemma. *K/I is a skew field.*

Proof. First note that since $M = U/I$ is an irreducible U-module, the ring $R = \mathrm{End}_U(M)$ by Schur's lemma is a skew field. Define a map $\varphi\colon K/I \to R$ by the rule $(\varphi(\bar{x}))\,(m) = (\varphi(x + I))\,(u + I) = ux + I$, where $m = u + I$, $\bar{x} = x + I$. It is easily seen that this definition is unambiguous and is an antihomomorphism of associative algebras. Moreover $\varphi(x + I)\,(1 + I) = x + I$, that is φ is injective. Further, if $\psi \in R$, then $\psi(1 + I) = y + I$ and $\psi(u + I) = \psi(u(1 + I)) = u\psi(1 + I) = uy + I \in U/I$. Thus $y \in K$ and $\psi = \varphi(y)$ so that φ is surjective. An obvious observation that a ring anti-isomorphic to a skew field is a skew field itself completes the lemma.

Note here that since A is central in U, $A \subseteq K$ and U/I is a module of finite type over a subring $A + I/I$ of a skew field K/I. To complete the proof of the theorem it suffices to prove that $A + I/I \cong A/A \cap I$ is a field of finite degree over k.

Indeed, let $a + I \neq I$ and $b + I \in K/I$ be such that $(a + I)\,(b + I) = 1 + I$. Consider a commutative ring $A[b]$; then U is an exact left $A[b]$-module, through left multiplication, of finite type over A. Using the Hamilton-Cayley theorem, the value in b of the characteristic polynomial of b in U acts as a zero linear transformation in U. Since U is exact over $A[b]$ this value itself equals zero, namely there exist $z_{n-1}, \dots, z_0 \in A$ such that $b^n + z_{n-1}b^{n-1} + \dots + z_0 = 0$, or

$$(b + I)^n + (z_{n-1} + I)\,(b + I)^{n-1} + \dots + (z_0 + I) = 0\,. \tag{1}$$

Multiplying both parts of (1) by $(a + I)^{n-1}$ one obtains

$$b + I = -\,(z_{n-1} + I) - \dots - (a + I)^{n-1}\,(z_0 + I) \in A + I/I\,,$$

which proves that $A + I/I$ is a field.

Now $A + I/I \cong A/A \cap I$ is a field finitely generated as an algebra over k by the images of $y_1, y_2, \dots, y_n$. Hence since by Hilbert's theorem it is well known that such a field is algebraic over k, $A + I/I$ is a finite extension of k; moreover, if k is algebraically closed, this extension is trivial.

This completes the proof of the theorem.

Corollary. *Let G be a finite dimensional Lie algebra over some field k of characteristic $p > 0$. Then $U(G)$ is a PI-algebra.*

Proof. Without any loss of generality only the case of an algebraically closed field may be considered. We denote by R the intersection of all kernels of irreducible representations of U and show that $R = \{0\}$. For, if $z \in R$ then $1 - z$ is invertible since otherwise $J = U(1 - z) \neq U$ and therefore is contained in some proper maximal left ideal I of U. Then U/I is a simple module so that $z \in I$; this implies, further, $1 \in I \neq U$. The latter contradiction shows that $1 - z$ is invertible in U. But an easy calculation using the fact that $\mathrm{gr}(U)$ is a polynomial ring shows that then $z = 0$.

Having proved this, $U(G)$ is a subdirect product of finite dimensional algebras U/I_α, I_α being a kernel of some irreducible representation. Since these dimensions are bounded by some number depending only on G, our corollary easily follows.

A more general version of this corollary will be proved in section 7, theorem 7.[1])

4.4. Some centralizer concerned techniques

Several sections to follow are devoted to finding of necessary and sufficient conditions for a Lie algebra G over some field k of prime characteristic $p > 0$ under which its universal envelope $U(G)$ is a PI-algebra. In this section we prove some auxiliary results related mainly with centralizers of elements in a Lie algebra (see definition in 1, 13). The first of these results is the following

Proposition 1. *Let M be a subset of a Lie algebra G over some ring k, $C_G(M) = \{x \in G \mid [x, M] = \{0\}\}$ be its centralizer in G. Then*

(i) *$C_G(M)$ is an a subalgebra of G;*

(ii) *if M is an ideal of G, then so is $C_G(M)$;*

(iii) *if $M = G^2$, the derived algebra of G, then $C_G(M)$ is nilpotent of class 2;*

(iv) *if M is an ideal of finite k-length, then so is $G/C_G(M)$.*

Proof. Denote $C_G(M)$ by C.

(i) Given $x, y \in C$, $m \in M$; using the Jacobi identity one has $[[x, y], m] = [[x, m], y] + [x, [y, m]] = 0$, that is $[x, y] \in C$.

(ii) Let $x \in G$, $y \in C$, $m \in M$ and $M \lhd G$. Then $[[x, y], m] = [[x, m], y] + [x, [y, m]] = 0$, since $[x, m] \in M$.

(iii) Let $x, y, z \in C$, $M = G^2$; then $[x, y, z] = 0$ for $x \in C$ and $[y, z] \in G^2 = M$.

(iv) Define $\varphi: G \to \mathrm{End}_k(M)$ by $\varphi(g)(m) = [g, m] \in M$. Then $\mathrm{Ker}\,\varphi = C$ so that by the homomorphism theorem G/C is isomorphic with a k-submodule of $\mathrm{End}_k(M)$ which has finite length together with M.

It follows from this theorem that denoting by $b(x)$ the *breadth* of the element $x \in G$, that is $b(x) = l(G/C_G(x))$, the breadths of elements of a Lie algebra G with $l(G^2) < \infty$ are finite and bounded by some number depending only on $l(G^2)$. Next we prove a converse result limited for the simplicity sake to the case of fields.

Theorem 3 (P. M. Neumann). *Let G be a Lie algebra over some field k such that $b(x) = \dim(G/C_G(x)) < b$ for some fixed b and any $x \in G$. Then $\dim(G^2) < b^2$.*

This theorem is in fact a theorem on bilinear maps and we formulate a proposition which will imply it. In this proposition given a bilinear map $\varphi: U \times V \to W$, we denote by $b_\varphi(u)$, $u \in U$, the codimension of the annihilator of u in V, that is $b_\varphi(u) = \dim(V/\mathrm{Ann}_\varphi(u)) = \dim(\varphi(u, V))$. Similarly, $b_\varphi(v) = \dim(U/\mathrm{Ann}(v)) = \dim(\varphi(U, v))$. As $\varphi(U, V)$ is not necessarily a subspace of W, we denote by $k\varphi(U, V)$ its linear span. Using these notations we have

Proposition 1. *Given linear spaces U, V, W over some field k and a bilinear map $\varphi: U \times V \to W$, assume that there exist positive integers r, s such that for all $u \in U$, $v \in V$, $b_\varphi(u) \leqq r$ and $b_\varphi(v) \leqq s$. Then $\dim(k\varphi(U, V)) \leqq rs$.*

[1]) An important corollary of the preceding argument is: if all irreducible G-modules have finite bounded dimension, then $U(G)$ is a PI-algebra.

Proof. Assume $r \leqq s$ and proceed by induction on r, the case $r = 0$ being trivial. So far if $b_\varphi(u) \leqq r - 1$ for all $u \in U$ then $\dim(k\varphi(U, V)) \leqq (r-1)s$. Suppose then for some $u_1 \in U$, $b_\varphi(u_1) = r$, and put $X = \mathrm{Ann}_\varphi(u_1) \subseteq V$, $Y = \mathrm{Ann}_\varphi(X) \subseteq U$. Let $\psi_1 \colon W \to W/\varphi(u_1, V) = W_1$ be a natural epimorphism and φ_1 a bilinear map $\varphi_1 = \psi_1\varphi \colon U \times V \to W_1$, the annihilator of u_1 under φ_1 coincides now with V. We need then a

Lemma. *If* $u \in U \setminus Y$, *then* $b_{\varphi_1}(u) \leqq r - 1$.

Indeed, given $u \notin Y$, there exists an $x \in X$ such that $\varphi(u, x) \neq 0$. Clearly, also $\varphi(u + u_1, x) \neq 0$. Denote by U^* the subspace $ku + ku_1 = k(u + u_1) + ku_1$, and by T the subspace $k\varphi(U^*, V)$. Then $T = \varphi(u, V) + \varphi(u_1, V) = \varphi(u, V) + \varphi(u + u_1, V)$. Since $\varphi(u, V) \cap \varphi(u + u_1, V) \ni \varphi(u, x) = \varphi(u + u_1, x) \neq 0$ and as $b_\varphi(u)$, $b_\varphi(u + u_1) \leqq r$, $\dim(T) \leqq 2r - 1$. Now $b_{\varphi_1}(u) = \dim \varphi_1(u, V) \leqq \dim(T/\varphi(u_1, V)) \leqq r - 1$ for by our assumption $b_{\varphi_1}(u_1) = \dim(\varphi(u_1, V)) = r$, q.e.d.

Returning to the proof of the proposition, note that, if $b_{\varphi_1}(y) \leqq r - 1$ for all $y \in Y$, then by the induction hypothesis $\dim(k\varphi_1(U, V)) \leqq (r-1)s$, so that $\dim(k\varphi(U, V)) = \dim(\varphi_1(U, V)) + \dim(\varphi(u_1, V)) \leqq (r-1)s + r \leqq (r-1)s + s = rs$ and we are finished.

Otherwise there exists a $u_2 \in Y$ such that $b_{\varphi_1}(u_2) = r$. Denote then by ψ_2 the natural epimorphism $W_1 \to W_1/\varphi_1(u_2, V) = W_2$. The composite map $\varphi_2 = \psi_2\varphi_1 \colon U \times V \to W_2$ is a bilinear map such that $\mathrm{Ann}_{\varphi_2}(u_2) = \mathrm{Ann}_{\varphi_2}(u_1) = V$. If $u \notin Y$, then $\varphi_2(u, V) = (\varphi_1(u, V) + \varphi_1(u_2, V))/\varphi_1(u_2, V)$, and the same argument as in the lemma shows that $b_{\varphi_2}(u) = \dim(\varphi_2(u, V)) \leqq r - 2$ for all $u \in U \setminus Y$.

Proceeding in the same manner, we obtain a sequence of bilinear maps $\varphi_1, \varphi_2, \dots, \varphi_t$, $t \leqq r$, such that (i) $\varphi_{i+1} = \psi_{i+1}\varphi_i$, $\dim(\mathrm{Ker}(\psi_i)) = r$, (ii) $b_{\varphi_i}(u) \leqq r - i$ for all $u \notin Y$, and (iii), $b_{\varphi_t}(y) \leqq r - t$, $y \in Y$, or $t = r$. If $t = r$, then $k\varphi_r(U, V) = k\varphi_r(Y, V)$ and $k\varphi(U, V) = k\varphi(Y, V)$ for $\mathrm{Ker}(\psi_r \dots \psi_1) \subseteq k\varphi(Y, V)$. But if $v_1, v_2, \dots, v_r$ span V modulo X, $k\varphi(Y, V) = \varphi(Y, v_1) + \dots + \varphi(Y, v_r) + \varphi(Y, X)$, whence remembering that $\dim(\varphi(Y, v_i)) \leqq s$ and $\varphi(Y, X) = \{0\}$, one has $\dim k\varphi(U, V) \leqq rs$.

If $t < r$, then by the induction hypothesis (since $b_{\varphi_t}(u) \leqq r - t$ for all $u \in U$), $\dim(k\varphi_t(U, V)) \leqq (r-t)s$. But $\dim(k\varphi(U, V)) = \dim(k\varphi_t(U, V)) + \dim(\mathrm{Ker}(\psi_t \dots \psi_1)) \leqq (r-t)s + tr \leqq (r-t)s + ts = rs$.

This completes the proof of the proposition 1 and the theorem 3.

A generalization of the situation of this theorem may be obtained as follows.

Let G be a Lie algebra over some field k, $\Delta_n = \Delta_n(G)$ be the set of all elements $x \in G$ such that $b(x) \leqq n$. Then though Δ_n is not even a subspace, we say that the elements $x_1, x_2, \dots, x_m$ are linearly dependent modulo Δ_n if there exist $\alpha_1, \alpha_2, \dots, \alpha_m \in k$ not all zero such that $\alpha_1 x_1 + \alpha_2 x_2 + \dots + \alpha_m x_m \in \Delta_n$.

Corollary. *Suppose for a Lie algebra* G *there exist positive integers* m, n *such that any* m *elements from* G *are linearly dependent modulo* Δ_n. *Then* G *possesses a subalgebra* H *such that* $\dim(G/H)$ *and* $\dim(H^2)$ *are finite.*

Proof. Let m be the least number with this property, and choose $e_1, e_2, \dots, e_{m-1}$ linearly independent modulo Δ_n. Denote by E the linear span of $\{e_1, e_2, \dots, e_{m-1}\}$ and by H the subalgebra generated by Δ_n. Then $H = \Delta_n + H \cap E$, since for any $h \in H$ one has $h + \alpha_1 e_1 + \alpha_2 e_2 + \dots + \alpha_{m-1} e_{m-1} \in \Delta_n$ by our conditions. Let $f_1, f_2, \dots, f_q$ be a basis of $H \cap E$. Remark, that if $x, y \in \Delta_n(G)$ then $x + y$ and $[x, y] \in \Delta_{2n}(G)$, for, say,

$C_G(x+y) \supseteq C_G(x) \cap C_G(y)$ and $\dim(G/C_G(x) \cap C_G(y)) \leqq 2n$. Using this remark one easily finds such a positive integer N that $b_G(h) \leqq N$ for any $h \in H$. Indeed, each element from H is representable as a linear combination of $\leqq m$ summands having the form $x \in \Delta_n$ or $\beta_i f_i$, $i = 1, 2, \ldots, q$. Each of these latter elements f_i lies in $H = \operatorname{alg}(\Delta_n)$ and as such an element is representable as a sum of finitely many monomials of some finite degree in Δ_n.

Now, clearly, $b_H(h) \leqq b_G(h) \leqq N$ for all $h \in H$. By theorem 3 then $\dim(H^2) \leqq N^2$, the inequality $\dim(G/H) < \infty$ following from $H \supseteq \Delta_n$. Thus the proof is complete.

And the last point we touch in this section is the

Proposition 2. *Let G be a Lie algebra over some field k and A be its abelian subalgebra of finite codimension. Then G possesses an abelian ideal H also having finite codimension in G.*

Proof. Denote by $[G^n, A] = [\underbrace{G, \ldots, G}_{n}, A]$ the subspace of G spanned by the monomials of the form $[g_1, g_2, \ldots, g_n, a]$, $g_i \in G$, $i = 1, 2, \ldots, n$, $a \in A$, $n \geqq 1$, and define a nondecreasing chain B_i, $i = 0, 1, 2, \ldots$, by $B_0 = A$, $B_{s+1} = B_s + [G^{s+1}, A]$. By our condition for some r, $B_{r+1} = B_r$ and B_r then is easily seen to be an ideal in G.

Let now $m = \dim(G/A)$ and let $g_1, g_2, \ldots, g_m$ form a basis of G modulo A. We define a nonincreasing chain of abelian subalgebras A_j, $j = 0, 1, 2, \ldots$, by putting $A_0 = A$ and taking for A_l as a subset of elements x from A_{l-1} such that $[x, g_j] \in A_{l-1}$, $j = 1, 2, \ldots, m$. Clearly, $[G, A_l] \subseteq A_{l-1}$. We show also that

(i) $\dim(G/A_l) < \infty$, $l = 0, 1, 2, \ldots$,

(ii) $[A_l, B_l] = \{0\}$, $l = 0, 1, 2, \ldots$,

using induction by l, the case $l = 0$ being trivial. Consider then the bilinear map $A_{l-1} \times G \to G \to G/A_{l-1}$ induced by the multiplication in G. Since A_l is a left kernel of this map and as $\dim(G/A_{l-1}) < \infty$ by the induction hypothesis, $\dim(A_{l-1}/A_l)$ is finite, the same then being true for $\dim(G/A_l)$. The assertion (ii) is as well simple. For, if already $[A_{l-1}, B_{l-1}] = \{0\}$, then by the Jacobi identity

$$[A_l, B_l] = [A_l, B_{l-1} + [G^l, A]] = [A_l, B_{l-1}] + [A_l, G^l, A] =$$
$$= [[A_l, G], [G^{l-1}, A]] + [G, A_l, G^{l-1}, A] \subseteq [A_{l-1}, B_{l-1}] = \{0\} .$$

Consider next the ideal B_r of the algebra G and denote by H its centre. Using (i) and (ii) one has

$$\dim(B_r/H) \leqq \dim(B_r/A_r) \leqq \dim(G/A_r) < \infty .$$

Now leaving to the reader the proof of the trivial fact that the centre of the ideal is itself an ideal we have completed the proof of the proposition.

4.5. Centralizers of elements in Lie algebras whose universal envelopes are PI-algebras

Let A be a ring which is a subring of a skew field Q. Call a skew field Q a *right quotient field* of A if each element from Q equals some fraction $a_1 a_2^{-1}$, $a_1, a_2 \in A$. We shall use in this section without proof the following theorem due to S. Amitsur ([64]):

Theorem 4. *Let A be an arbitrary PI-algebra without zero divisors over a commutative field k. Then A possesses a right quotient field Q finite dimensional over its centre.*

This theorem is basic for the next

Proposition 1. *Let G be a Lie algebra over some commutative field k, $U(G)$ be its universal envelope satisfying a nontrivial polynomial identity. Then there exist positive integers m and n such that any m elements of G are linearly independent modulo $\Delta_n(G)$.*

Proof. Since $U(G)$ has no zero divisors the theorem 4 applies. Denote the dimension of a quotient field Q of $U = U(G)$ over its centre Z by $m - 1$. Then, given any m elements $x_1, x_2, \dots, x_m$ from G there exist central elements $a_1b_1^{-1}, a_2b_2^{-1}, \dots, a_mb_m^{-1}$, $a_i, b_i \in U$, $1 \leqq i \leqq m$, not all zero, such that

$$x_1a_1b_1^{-1} + x_2a_2b_2^{-1} + \dots + x_ma_mb_m^{-1} = 0\,. \tag{1}$$

Let x be any element from G; commuting (1) with x, multiplying then both sides by $b_1b_2 \dots b_m$ and noting that $a_ib_i^{-1}b_1b_2 \dots b_m = b_1 \dots b_{i-1}a_ib_i^{-1}b_i \dots b_m = c_i \in U$, $i = 1, 2, \dots, m$, one obtains

$$[x_1, x]\,c_1 + [x_2, x]\,c_2 + \dots + [x_m, x]\,c_m = 0\,,$$

where not all c_i equal zero. Consider $R = \mathrm{gr}(U) = k[G]$, where G is identified with a subspace of elements of degree 1 in a polynomial ring $k[G]$, and denote by $f_1, f_2, \dots, f_m$ the images of $c_1, c_2, \dots, c_m$ in R. Then we have an equality

$$[x_1, x]\,f_1 + [x_2, x]\,f_2 + \dots + [x_m, x]\,f_m = 0 \tag{2}$$

in $R = k[G]$ satisfied identically by $x \in G$.

Now we may consider a system

$$a_{x1}t_1 + a_{x2}t_2 + \dots + a_{xm}t_m = 0\,, \qquad x \in G\,, \qquad a_{xi} = [x, x_i]\,, \tag{3}$$

of linear equations with coefficients in a polynomial ring $R = k[G]$ having a nontrivial solution $(f_1, f_2, \dots, f_m)$ from (2). Then since R has no zero divisors, there exists some l, $0 \leqq l \leqq m$, such that all minors of the matrix of the coefficients of (3) having order $> l$ equal zero. Choose l minimal with this property. If $l = 0$, then $a_{xi} = [x, x_i] = 0$ for all $x \in G$, whence $x_i \in Z(G) \subseteq \Delta_n$, $n \geqq 0$. If $l > 0$ then there is a nontrivial minor P of order l with entries $a_{y_si_r}, y_1, y_2, \dots, y_l \in G$, $\{i_1, i_2, \dots, i_l\} \subset \{1, 2, \dots, m\}$. Let $i_{l+1} \in \{1, 2, \dots, m\} \setminus \{i_1, i_2, \dots, i_l\}$, x be an arbitrary element from G. Then decomposing the following minor of order $l + 1$

$$0 = \begin{vmatrix} \quad P \quad & \vdots\, a_{y_1i_{l+1}} \\ \dots\dots\dots & \vdots\, a_{y_li_{l+1}} \\ a_{xi_1}\, a_{xi_2} \,\cdots\, a_{xi_l} & a_{xi_{l+1}} \end{vmatrix} = \begin{vmatrix} [y_1, x_{i_1}] \dots [y_1, x_{i_l}] & [y_1, x_{i_{l+1}}] \\ [y_l, x_{i_1}] \dots [y_l, x_{i_l}] & [y_l, x_{i_{l+1}}] \\ [x, x_{i_1}] \dots [x, x_{i_l}] & [x, x_{i_{l+1}}] \end{vmatrix}$$

by the last row we obtain a relation of the form

$$[x_1, x]\,g_1 + [x_2, x]\,g_2 + \dots + [x_m, x]\,g_m = 0\,, \tag{4}$$

where not all of $g_1, g_2, \dots, g_m$ equal zero (e.g. $g_{i_{l+1}} = P \neq 0$). Moreover if V is a subspace of G spanned by the elements $[y_s, x_{i_r}]$, $1 \leqq s \leqq l$, $1 \leqq r \leqq l + 1$, then $\dim(V) \leqq l(l + 1) < m^2$ and $g_1, g_2, \dots, g_m \in k[V]$.

We put now $n = m^2$ and, assuming without loss of generality that $g_m \neq 0$, prove that $x_1, x_2, \dots, x_m$ are linearly dependent modulo $\Delta_n(G)$. Use induction on m, the number of nonzero coefficients $g_1, \dots, g_m$ in (4). If $m = 1$, then $[x_m, x]\,g_m = 0$ implies in R, having no zero divisors, that $[x_m, x] = 0$ for all $x \in G$, that is, $x_m \in Z(G) \subseteq \Delta_n$, $n \geqq 1$. Let then $m > 1$, and e.g., $g_1 \neq 0$. Suppose first that for some i, $[x_i, x] \in V$ for all $x \in G$. Then $[x_i, G] \subseteq V$ and $b(x_i) = \dim(G/C_G(x_i)) \leqq \dim V < m^2$, so that $x_i \in \Delta_n$ and we are finished. Let second there exist a $y_0 \in G$ such that $[x_1, y_0] = e \notin V$. Choose a linear basis in G including that of V and e, and put $[x_i, y_0] = \alpha_i e + h_i$, $i = 2, 3, \ldots, m$; h_i written in the elements of this basis but e. Putting y_0 into (4) one obtains

$$e(g_1 + \alpha_2 g_2 + \ldots + \alpha_m g_m) + h_2 g_2 + \ldots + h_m g_m = 0 .$$

Using the standard basis of a polynomial ring, this equality implies

$$g_1 + \alpha_2 g_2 + \ldots + \alpha_m g_m = 0$$

which transforms (4) into

$$[x_2 - \alpha_2 x_1, x]\, g_2 + [x_3 - \alpha_3 x_1, x]\, g_3 + \ldots + [x_m - \alpha_m x_1, x]\, g_m = 0$$

so that the induction hypothesis applies. Noting that a nontrivial dependence of elements $x_2 - \alpha_1 x_1, \ldots, x_m - \alpha_m x_1$ modulo Δ_n yields that of $x_1, x_2, \ldots, x_m$, we complete the proof of the proposition.

Corollary. *Under the same conditions G possesses a subalgebra H with* $\dim(H^2) < \infty$ *such that* $\dim(G/H) < \infty$.

This follows immediately from the just proved proposition and the corollary of theorem 4.3.

Our aim in the second part of this section is the

Theorem 5. *Let G be a Lie algebra over some field k and assume that its universal envelope $U(G)$ is a PI-algebra. Then G possesses an abelian ideal of finite codimension bounded by a number depending on the degree of a nontrivial identity holding in $U(G)$.*

We start with an example of a Lie algebra whose universal envelope is not PI.

Example. Let A be a nilpotent Lie algebra of nilpotency class two having basis $\{x_1, x_2, \ldots, x_{d+1}, y_1, y_2, \ldots, y_{d+1}, z\}$ and the multiplication table $[x_i, y_j] = \delta_{ij} z$, $1 \leqq i, j \leqq d + 1$, δ_{ij} being Kronecker's delta, $[x_i, x_j] = [y_i, y_j] = [x_i, z] = [y_j, z] = 0$. Show that $U(A)$ satisfies no polynomial identity of degree $\leqq d$. For, let on the contrary

$$X_1 X_2 \ldots X_d + \sum_{\sigma \neq 1} \alpha_\sigma X_{\sigma(1)} X_{\sigma(2)} \ldots X_{\sigma(d)} \equiv 0 \tag{5}$$

be a nontrivial multilinear identity of the degree d, holding in $U(A)$. Order the basis of A so that

$$x_1 > x_2 > \ldots > x_{d+1} > y_1 > y_2 \ldots > y_{d+1} > z .$$

Then by the PBW-theorem 2,4.6 the basis of $U(A)$ is formed the by monomials

$$x_1^{s_1} x_2^{s_2} \ldots x_{d+1}^{s_{d+1}} y_1^{r_1} y_2^{r_2} \ldots y_{d+1}^{r_{d+1}} , \qquad s_i, r_j, t \geqq 0 . \tag{6}$$

To calculate the value of the left hand side of (5) in $X_1 = x_1 y_2$, $X_2 = x_2 y_3, \ldots, X_d = x_d y_{d+1}$, we need the following argument.

Let $Q = (q_{ij})$ be an upper triangular $d \times d$ matrix each column of which contains exactly one nonzero element equal to1. Then for each substitution $\delta \in S_n$ the element

$$x_{\sigma(1)} y_{\sigma(1)+1} x_{\sigma(2)} y_{\sigma(2)+1} \ldots x_{\sigma(d)} y_{\sigma(d)+1}$$

can be represented as a sum

$$\sum_Q \pm x_{\sigma(1)}^{q_{11}} x_{\sigma(2)}^{q_{12}} \dots x_{\sigma(d)}^{q_{1d}} \big((\mathrm{ad}(x_{\sigma(2)}))^{q_{22}} \dots (\mathrm{ad}(x_{\sigma(d)}))^{q_{2d}} (y_{\sigma(1)+1})\big) \dots \big((\mathrm{ad}(x_{\sigma(d)}))^{q_{dd}} (y_{\sigma(d)+1})\big) \tag{7}$$

where the summation is taken over all described matrices Q.

This can be proved by direct calculations using induction on d and having in mind that $\big(\mathrm{ad}(x_i)\big)\ (y_j)$ is central.

Each summand in (7) is either 0 or equal to a basic element of $U(A)$. Consider a set of those summands which have the form (6) with $t = d - 1$. Obviously such a monomial arises only when Q is a unity matrix, that is $q_{ii} = 1$ for all $1 \leqq i \leqq d$. This single monomial is

$$\pm\, x_{\sigma(1)} \big(\mathrm{ad}(x_{\sigma(2)})\big)\ (y_{\sigma(1)+1}) \dots \big(\mathrm{ad}(x_{\sigma(d)})\big)\ (y_{\sigma(d-1)+1})\ y_{\sigma(d)+1}\ . \tag{8}$$

If $\sigma = 1$, (8) equals $x_1 y_{d+1} z^{d-1}$, otherwise $\big(\mathrm{ad}(x_{\sigma(j)})\big)\ (y_{\sigma(j-1)+1}) = [x_{\sigma(j)}, y_{\sigma(j-1)+1}] = 0$ for some j, $\sigma(j) \neq \sigma(j-1) + 1$.

From this argument it follows that the above mentioned value of the left hand side of (5) is not zero, which contradicts our assumption. To obtain thus an example of a class 2 nilpotent algebra whose universal envelope is not a PI, it suffices to consider an algebra with the same multiplication table as A but with infinitely many basic elements x_i, y_j.

Now we return to the proof of the theorem 5.

Proof. By the corollary preceding the present theorem it suffices to limit oneself to the case of an algebra H with $\dim (H^2) < \infty$. For, it is an easy exercise following from the PBW-theorem 2,4.6 that if H is a subalgebra of G then $U(H)$ is a subalgebra of $U(G)$ and hence is a PI-algebra. Now suppose we have proved our theorem for H and found in H an abelian ideal K of finite codimension. Then K is an abelian subalgebra of finite codimension in G and it remains to apply proposition 4.2.

So far, consider $C = C_H(H^2)$. By proposition 4.1, (ii) and (iii), C is an ideal of finite codimension in H and of nilpotency class 2. By the same reason as before, it suffices to prove that the centre of C, $Z(C) = Z$, has finite codimension in C. Moreover, further reduction may be obtained as follows. Since $\dim(C^2) = t < \infty$ and $C^2 \subseteq Z(C)$, represent $\{0\}$ as an intersection of subspaces D_i, $i = 1, 2, \dots, t$, $\dim C^2/D_i = 1$. Then C is a *subdirect product* of algebras $C_i = C/D_i$, such that $\dim(C_i^2) = 1$. In other words, the map $\varphi\colon C \to \Pi\, C_i$ given by $\varphi(c) = (c + D_1, c + D_2, \dots, c + D_t)$ is a monomorphism such that the projection of $\varphi(C)$ onto C_i equals this latter algebra. Clearly $U(C_i)$ is a PI-algebra (as a homomorphic image of $U(C)$) and our purpose is to show that its centre has finite codimension in C_i.

But since C_i^2 is one dimensional and spanned, say, by e, the multiplication in C_i gives an alternating bilinear form $(x, y) = \alpha$ if $[x, y] = \alpha e$. Such a form is well known to have a canonical basis $\{x_1, \dots, x_s, \dots; y_1, \dots, y_r, \dots; z_1, z_2, \dots\}$ such that $-(y_r, x_s) = (x_s, y_r) = \delta_{sr}$, $(x_i, x_j) = (y_i, y_j) = (x_i, z_j) = (y_i, z_j) = (z_i, z_j) = 0$.

If the centre of C_i had infinite codimension, then for any $d > 0$, C_i contained a subalgebra isomorphic to the algebra of the example preceding the present proof, and we see that $U(C_i)$ cannot satisfy a nontrivial polynomial identity. This contradiction completes the proof of the theorem.

4.6. Inner derivations in Lie algebras whose universal envelopes are PI-algebras

Let V be a vector space over some field k, φ be a linear transformation of V. We say that φ is *algebraic of degree* $\leqq d$ if φ is a root of a polynomial f of degree d with coefficients in k.

Theorem 6. *Let G be a Lie algebra over a commutative field k of prime characteristic such that $U(G)$ is a PI-algebra. Then, for any $x \in G$,* $\mathrm{ad}(x)$ *is algebraic of degree $\leqq N$, N depending only on G.*

The **proof** of this theorem is rather combinatorial. Suppose that $U(G)$ is a PI-alge- r asatisfying a mu ltilinear identity

$$P(x_1, x_2, \dots, x_d) = \sum_{\sigma \in S_d} \alpha_\sigma x_{\sigma(1)} x_{\sigma(2)} \dots x_{\sigma(d)} \equiv 0\,, \tag{1}$$

S_d being the group of all substitutions of $\{1, 2, \dots, d\}$, $\alpha_\sigma \in k$, $\alpha_1 \neq 0$.

Assume first that there are elements v and g in G such that the elements $v_0 = v$, $v_1 = (\mathrm{ad}(g))\,(v), \dots, v_{N+1} = (\mathrm{ad}(g))^{N+1}\,(v)$ are linearly independent, $N = 1 + p + p^2 + \dots + p^{2d}$. Obviously, it follows then that the system $g, v_1, \dots, v_N$ is linearly independent. Include this system as terminal segment into some totally ordered basis of G. Then the basis of $U(G)$ will be formed by monomials of the form

$$g^n v_0^{n_0} v_1^{n_1} \dots v_N^{n_N} e_1^{l_1} e_2^{l_2} \dots e_t^{l_t} \dots\,, \tag{2}$$

where $e_1 \geqq e_2 \dots$ and all but a finite number of nonnegative integers $n, n_0, n_1, \dots, n_N$, $l_1, l_2, \dots, l_t, \dots$ equal zero.

We show now that the element

$$P(g^d v_{p^d+1}, g^{p^2} v_{p^d+2}, \dots, g^{p^d} v_{p^{2d}}) \tag{3}$$

is a nonzero element of $U(G)$ contradicting (1). Indeed, (3) is a linear combination of monomials of the form

$$B_\sigma = g^{p^{\sigma(1)}} v_{p^d+\sigma(1)} g^{p^{\sigma(2)}} v_{p^d+\sigma(2)} \dots g^{p^{\sigma(d)}} v_{p^d+\sigma(d)}\,. \tag{4}$$

Lemma. *Let $Q = (q_{ij})$ be an upper triangular $d \times d$ matrix such that all but one element of each column are zero, and this one element equals 1. Then B_σ from (4) can be represented in the form*

$$\sum_Q \pm\, g^{p^{\sigma(1)} + q_{12} p^{\sigma(2)} + \dots + q_{1d} p^{\sigma(d)}} v_{p^d+\sigma(1) + q_{22} p^{\sigma(2)} + \dots + p_{2d} p^{\sigma(d)}} \dots$$

$$\dots v_{p^d+\sigma(d-1) + q_{dd} p^{\sigma(d)}} v_{p^d+\sigma(d)}\,, \tag{5}$$

where the sum is taken over all possible matrices of the described form.

Induction by d, the lemma being trivial for $d = 1$. The general case is dealt with by transforming of the initial segment of (4) without $g^{p^{\sigma(d)}} v_{p^d+\sigma(d)}$ using the induction hypothesis and then permuting $g^{p^{\sigma(d)}}$ with the 'v-tail' of (5) using that fact that bracketing is derivation and that $(\mathrm{ad}(g^p))\,(v_q) = (\mathrm{ad}(g))^p\,(v_q) = v_{p+q}$. Routine calculations are left to the reader.

Now clearly by our condition the elements

$$v_{p^d+s+\varepsilon_1 p+\varepsilon_2 p^2+\dots+\varepsilon_d p^d}\,, \qquad 1 \leqq s \leqq d\,, \qquad \varepsilon_r = 0, 1\,, \qquad r = 1, 2, \dots, d\,, \tag{6}$$

are different elements of the chosen basis of G for different sequences $(s, \varepsilon_1, \varepsilon_2, \ldots, \varepsilon_d)$. Denoting by $\overline{M}$ the element of the highest degree in the representation of some monomial M from (5) in (2), one observes, that $\overline{M}$ differs from M only by the order of 'v-factors', that is factors of the form (6).

So far, choose from (5) a single monomial C_σ in which only one 'v-factor' has the form (6) with $\varepsilon_1 = \varepsilon_2 = \ldots = \varepsilon_d = 0$. Clearly this monomial has the form

$$C_\sigma = g^{p^{\sigma(1)}} v_{pd+\sigma(1)+p^{\sigma(2)}} \, v_{pd+\sigma(2)+p^{\sigma(3)}} \cdots v_{pd+\sigma(d)} \tag{7}$$

and is linearly independent from the set of all other monomials in (5). Moreover, it is easily seen that (7) may be linearly dependent only on a monomial of the same form for other σ.

Assume then that $C_1 = C_\sigma$ for some σ. Then clearly $p^{\sigma(1)} = p^1$, from which $\sigma(1) = 1$. Whence $v_{pd+\sigma(1)+p^{\sigma(2)}} = v_{pd+1+p^2}$, and therefore $\sigma(2) = 2$. Going on similarly, we get $\sigma = 1$. So C_1 is linearly independent on all other summands of (3) and has a coefficient $\alpha_1 \neq 0$.

The obtained contradiction shows that given $g \in G$, for any $v \in G$ there exists a polynomial $\varphi_v(t) \in k[t]$ such that $(\varphi_v(\mathrm{ad}(g)))\,(v) = 0$ and $\deg(\varphi_v) \leq N + 1$. Consider then G as a periodic $k[t]$-module through $tv = (\mathrm{ad}(g))\,(v)$ and decompose G in the direct sum of primary components $\bigoplus_\mu G_\mu$, μ being an irreducible polynomial from $k[t]$. Obviously at most $N + 1$ summands G_μ may be nonzero, for otherwise an element $v = v_1 + v_2 + \ldots + v_{N+2}$, $0 \neq v_i \in G_{\mu_i}$, $i = 1, 2, \ldots, N + 2$ — is not annihilated by any polynomial of degree $\leq N + 1$. Now since each of $G_{\mu_1}, G_{\mu_2}, \ldots, G_{\mu_{N+1}}$ is annihilated by μ_i^{N+1}, the derivation $\mathrm{ad}(g)$ is annihilated by $(\mu_1 \ldots \mu_{N+1})^{N+1}$ and hence is algebraic of bounded degree. This proves the theorem.

4.7. Main theorem on polynomial identities in universal envelopes of Lie algebras

In this section we give necessary and sufficient conditions for a Lie algebra G over some field of characteristic $p > 0$ which ensure $U(G)$ to be a PI-algebra.

Theorem 7 (Bahturin). *Let G be a Lie algebra over some commutative field of prime characteristic $p > 0$. Then its universal envelope $U = U(G)$ is a PI-algebra if and only if the following two conditions hold:*

(i) *G contains an abelian ideal H such that $\dim(G/H) < \infty$;*

(ii) *for any $x \in G$, $\mathrm{ad}(x)$ is algebraic of bounded degree $\leq N$, depending only on G.*

Proof. The necessity of these conditions having been proved in theorems 5.5 and 6.6, it remains to prove that these conditions are sufficient.

Thus represent G as a direct sum of H and some complementary subspace $F: G = H \oplus F$ and choose a totally ordered basis E in H and $f_1, f_2, \ldots, f_n$ in F, $n = \dim(G/H)$. Then by the PBW-theorem 2,4.6 one may choose a basis in $U(G)$ of the form

$$e_1 e_2 \ldots e_t f_1^{r_1} f_2^{r_2} \ldots f_n^{r_n}, \quad e_1 \geq e_2 \geq \ldots \geq e_t, \quad e_i \in E, \quad r_j \geq 0.$$

Just as in the case of lemma 1.1, using that $\mathrm{ad}(f_i), (\mathrm{ad}(f_i))^2, \ldots, (\mathrm{ad}(f_i))^N$ by our condition are linearly dependent, one may choose such 'p-polynomials' $\varphi_1, \varphi_2, \ldots, \varphi_n$ that

$\varphi_i(\mathrm{ad}(f_i)) = 0$, and $z_i = \varphi_i(f_i) \in Z(U)$, $i = 1, 2, \ldots, n$. Then as in proposition 1.2 a basis of U will be formed by the monomials

$$e_1 e_2 \ldots e_t z_1^{s_1} z_2^{s_2} \ldots z_n^{s_n} f_1^{r_1} f_2^{r_2} \ldots f_n^{r_n}, \tag{1}$$
$$e_1 \geqq \ldots \geqq e_t, \quad e_i \in E, \quad s_j \geqq 0, \quad 0 \leqq r_j < \deg \varphi_j.$$

It is easily seen that the elements (1) with $r_1 = r_2 = \ldots = r_n = 0$ form a commutative subring S of U (a polynomial ring $k[E; z_1, z_2, \ldots, z_n]$) and that U is a free left S-module of finite type $\leqq (\deg \varphi_1) \ldots (\deg \varphi_n)$. Consider $\mathrm{End}_S(U)$ and imbed U into $\mathrm{End}_S(U)$ by the map $x \mapsto R_x$, $x \in U$. ($R_x \in \mathrm{End}_S(U)$ since U is a left S-module and R_x is the multiplication from the right.) But by the final remark from section 1 of this chapter $\mathrm{End}_S(U)$ satisfies a standard identity and thus is a PI-algebra. The same is hence true for U, and the proof of the theorem is complete.

4.8. Residual properties of Lie algebras

In the remaining part of this chapter we discuss miscellaneous *finiteness conditions* for Lie algebras, that is, those properties of infinite dimensional Lie algebras which trivially hold in finite dimensional Lie algebras. The examples of such conditions are numerous and we start with so called *residual properties*. Let $\mathfrak{X}$ be a class of Lie algebras over some ring k and G be an algebra over k. We say that G is *'residually an algebra from* $\mathfrak{X}$', or simple *'residually* $\mathfrak{X}$', if for each $x \neq 0$ from G there exists an algebra $A_x \in \mathfrak{X}$ and an epimorphism $\varphi_x: G \to A_x$ such that $\varphi_x(x) \neq 0$. Let, in particular, $\mathfrak{F}$ be the class of all finite length algebras over k; then a residually $\mathfrak{F}$-algebra will be called *residually finite*. Any abelian algebra A over a field k is easily seen to be residually finite, on the other hand an abelian **Z**-Lie algebra whose underlaying group is the group U_{p^∞} of all complex roots of 1 having p-power degree, p being prime, or the additive group of rational numbers are not residually finite.

To give some necessary and sufficient conditions for the residual finiteness of a Lie algebra we need the concept of a *monolithic* Lie algebra. Assume that in a Lie algebra A over some ring k the intersection M of all nonzero ideals is nonzero. Then M is called a *monolith* of A and A is called *monolithic*. Recall also that the algebra A is called a *subdirect product* (see also the proof of theorem 5.5) of a set A_α, $\alpha \in I$, of Lie algebras if A is isomorphic to a subalgebra A' of a cartesian product $P = C\prod_{\alpha \in I} A_\alpha$ such that $\pi_\alpha(A') = A_\alpha$, π_α being a projection of P onto A_α, $\alpha \in I$.

Proposition 1. (i) *Any Lie algebra A is isomorphic to a subdirect product of its monolithic homomorphic images;*

(ii) *A monolithic Lie algebra is residually finite if and only if it is of finite length over k;*

(iii) *A Lie algebra A is residually $\mathfrak{X}$, $\mathfrak{X}$ being some class of Lie algebras over k, if and only if A is a subdirect product of algebras from $\mathfrak{X}$.*

Proof. (i) Assume A non-monolithic. Let $x \neq y$ and, say, $y \neq 0$ be elements from A. Consider a maximal ideal $I_{x,y}$ of A such that $I_{x,y} \ni x$ and $I_{x,y} \not\ni y$ existing by Zorn's lemma. Clearly by our choice any nonzero ideal of $A/I_{x,y}$ contains $y + I_{x,y} \neq 0$, that is $A/I_{x,y}$ is monolithic. Moreover $\bigcap_{x,y} I_{x,y} = \{0\}$ since already $\bigcap_{y \neq 0} I_{0,y} = \{0\}$. Put then $P = C \prod_{\substack{x \neq y \neq 0 \\ x,y \in A}} (A/I_{x,y})$ and for each $z \in A$ define $\varphi_z \in P$ as a function whose value in $A/I_{x,y}$

equals $z + I_{x,y}$. Obviously $z \mapsto \varphi_z$ is a monomorphism and under this monomorphism A is a subdirect product in P.

(ii) Clearly only the necessity of the condition needs verification. Thus assume that A is monolithic with the monolith $M \neq \{0\}$ and is of infinite length over k. Take $x \in M \setminus \{0\}$. Then the only ideal of A not containing x is $\{0\}$ and therefore by the homomorphism theorem x lies in the kernel of any homomorphism of A onto any algebra of finite length over k.

(iii) Remark that if A is residually $\mathfrak{X}$ then A contains a set $\{J_\alpha, \alpha \in I\}$ of ideals with $\bigcap_{\alpha \in I} J_\alpha = \{0\}$ and $A/J_\alpha \in \mathfrak{X}$ for all $\alpha \in I$. Just as in (i) A is then isomorphic with a subdirect product of A/J_α, $\alpha \in I$. The converse assertion is also very simple.

Note that though if all monolithic images of A are of finite length then by (i) and (ii) A is residually finite, the converse is wrong. For example, any free Lie algebra is residually finite, but all monolithic algebras are the quotient algebras of this latter.

Next we give three important examples of non residually finite algebras. There main source is part (ii) of the just proved proposition.

Examples. I. Let A be a Lie algebra analogous to that of the example from section 5 but infinite dimensional over some field k. Recall its basis $\{x_i, y_j, z\}$, where now i, j runs through all integers, and the multiplication table $[x_i, y_j] = \delta_{ij} z$, $[x_i, x_j] = [y_i, y_j] = [x_i, z] = [y_j, z] = 0$. Clearly, A is monolithic with monolith $M = kz$ spanned by z. By 1 (ii) A is not residually finite. Thus we have an example of an infinitely generated algebra over a field of nilpotency class 2 which is not residually finite.

II. Now we construct an example of a finitely generated Lie algebra with some additional properties which is not residually finite. Namely, let k be a field of characteristic 0 and let $P = A \leftthreetimes M$ be as in the example (i), section 1, 5. Then since M is an irreducible A-module, one immediately observes that M is a monolith of P and thus by 1 (ii) P is not residually finite. On the other hand, given any nonzero element $f \in M$, P is generated by e_1, e_2, f. Hence we have an example of a finitely generated extension of a class 2 nilpotent Lie algebra which is not residually finite.

III. More elaborate is the following example: consider a Lie algebra A from the example I and define a linear map $d\colon A \to A$ by $d(x_i) = x_{i+1} - x_{i-1}$, $d(y_i) = y_{i+1} - y_{i-1}$, $d(z) = 0$. One immediately verifies that d is a derivation of A, that is $D = kd \subseteq \mathrm{Der}(A)$. Form next a semidirect product $T = D \leftthreetimes A$. Then T is generated, say, by d, x_0, x_1, y_0, y_1 and not residually finite since $T \supseteq A$ while the residual finiteness is a property inherited by subalgebras. Remark also that K belongs to a class of so called *'centre-by-metabelian'* Lie algebras, that is $K'' = (K^2)^2 \subseteq Z(K)$. This can be equivalently defined by the identity $[v, [x, y], [z, u]] \equiv 0$.

After these examples the main theorem of this section seems to be of certain interest.

Theorem 8 (BAHTURIN). (1) *Let k be any field, A be some finitely generated metabelian algebra over k. Then A is residually finite.*

(2) *Let k be a field of prime characteristic $p > 0$, A be some finitely generated algebra over k possessing an abelian ideal B such that* $\dim(A/B) < \infty$. *Then A is residually finite.*

Lemma. *Let G be a finite dimensional Lie algebra over some field k. Then its universal envelope $U = U(G)$ is a noetherian algebra, that is, satisfies maximality conditions both for left and right ideals.*

Proof. Consider $\overline{U} = \operatorname{gr}(U)$. By the corollary from 2,4, given a basis $e_1, e_2, \dots, e_n$ in G, $\overline{U} = k[e_1, e_2, \dots, e_n]$ which is a noetherian ring. To prove the maximality conditions for U it suffices to show that any, say, left ideal I of U is generated by a finite number of elements. Let then $\overline{I}$ be the set of all major parts $\overline{a}$ of elements a from I, that is, elements of the form $a + U_m$, where $a \in U_{m+1} \setminus U_m$. The reader immediately observes that $\overline{I}$ is an ideal of $\overline{U}$ having thus a finite number of generatoıs $\overline{a}_1, \overline{a}_2, \dots, \overline{a}_r \in \overline{I}$. Using induction by m introduced above, we show that a is an element of the ideal of U generated by $a_1, a_2, \dots, a_r$. For, since $\overline{U}$ is homogeneous, $\overline{a} \in \overline{I}$, $\overline{a} = \sum_{i=1}^{r} \overline{b}_i\overline{a}_i$, where all $\overline{b}_i\overline{a}_i \in \overline{U}_{m+1}$, and $a' = a - \sum_{i=1}^{r} b_i a_i$ is an element of $U_m \cap I$. Now the induction hypothesis finishes the proof of the lemma.

In both cases, using proposition 1 (ii), it suffices to prove that finitely generated monolithic algebras with the same properties are finite dimensional. Remark that in the case (1) A also possesses an abelian ideal B of finite codimension in A, for example, A^2, the derived algebra of A, is such an ideal. Denote then A/B by G and consider a universal envelope $U = U(G)$ of G. We prove that B is a U-module of finite type and that M, the monolith of A, is an irreducible submodule of this module.

Recall that (cf. section 3 or 2, 4) the U-module structure in B can be obtained from the G-module one using the universal property of U: for given G-module structure in B, there exists a homomorphism of Lie algebras $\varphi\colon G \to \operatorname{End}_k(B)$ and hence a unique homomorphism of associative algebras $\psi\colon U \to \operatorname{End}_k(B)$ extending φ, endowing thus B by a U-module structure. (Recall also (see 1, 15) that the G-module structure of B is given by $gb = (a + B)\, b = [a, b]$, if $g = a + B$.) To prove that B is a finitely generated U-module, represent A as a quotient algebra L/N of some free Lie algebra L of finite rank by an ideal N, B being then represented as a quotient algebra $B = B'/N$, where L/B' is finite dimensional. Let $e_1, e_2, \dots, e_s$ be a basis of L modulo B'. Then for any $1 \leqq i, j \leqq s$, there exist elements $c_{ij}^{\nu} \in k$ such that $r_{ij} = [e_i, e_j] - \sum_{\nu=1}^{s} c_{ij}^{\nu} e_\nu \in B'$. Evidently then B' as an ideal of L is generated by r_{ij}, $1 \leqq i, j \leqq n$ and therefore $B = B'/N$ is finitely generated as an ideal of $A = L/N$. Since further B is abelian, it is finitely generated as an $A/B = G$-module, whence as a U-module. To complete our proof it is sufficient to show that B is finite dimensional over k. By the previous remark B is the sum of finitely many cyclic U-submodules and therefore it remains to consider only the case of cyclic monolithic (that is with the least nonzero submodule) U-module B with monolith M.

Since G may be chosen abelian in the case (1), U is either a polynomial algebra in finitely many variables or a universal envelope of a finite dimensional algebra G over a field k of prime characteristic $p > 0$. In both cases its irreducible modules are known to be finite dimensional (see theorem 3.2 and the final paragraph of its proof where it is noted that given a maximal ideal I of $k[y_1, y_2, \dots, y_n]$, the quotient module $k[y_1, y_2, \dots, y_n]/I$ is finite dimensional over k). Again in both cases U possesses a central subalgebra $R = k[y_1, y_2, \dots, y_n]$ such that U is a free R-module of finite type. For, in the first case $U = R$, and in the second case lemma 1.1 applies.

So far, let $I = \operatorname{Ann}_U(a_0)$ be the annihilator of some nonzero element a_0 of the least submodule M in U. Since M is irreducible, $U/I \cong M$ as a U-module, so that U/I

and therefore $R/R \cap I$ are finite dimensional over k. Since $R \cap I$ is central in U, clearly $R \cap I \subseteq \mathrm{Ann}_U(M)$, the annihilator of M in U. Choose then $\zeta \in R \cap I$ and consider a nonincreasing chain $B_i = \zeta^i B$, $i = 0, 1, \ldots$, of U-submodules in B. Clearly, either for some s, $B_s = \{0\}$, or for all i, $i = 0, 1, 2, \ldots$, $B_i \supseteq M$. Let $\varphi_i \colon x \mapsto \zeta^i x$ be a mapping of B into B_i; then φ_i is a homomorphism of U-modules and therefore its kernel C_i is a U-submodule in B, which contains M, since $\zeta \in \mathrm{Ann}_U(M)$. By the lemma U is a noetherian ring and hence B is a noetherian U-module. In this latter we have a nondecreasing chain C_i, $i = 1, 2, \ldots$, of U-submodules and therefore there exists a positive integer $r \geqq 1$ such that $C_r = C_{r+1} = \ldots$. If $B_r \neq \{0\}$, then $B_r \supseteq M$ and taking $a \neq 0$ in M, there exists a $b \in B$ such that $\varphi_r(b) = \zeta^r b = a \neq 0$ but $\varphi_{r+1}(b) = \zeta^{r+1} b = \zeta a = 0$, so that $b \in C_{r+1} \backslash C_r$ which is impossible. Thus we see that for any $\zeta \in R \cap I$ one has $\zeta^r B = \{0\}$ for some positive integer r.

Now we are in a position to prove that B is finite dimensional. For, let a be a generator of the cyclic $U = U(G)$-module B. Using lemma 1.1, given a basis $e_1, e_2, \ldots, e_n$ of G, choose a basis of U in the form

$$e_1^{l_1} e_2^{l_2} \ldots e_n^{l_n} y_1^{s_1} y_2^{s_2} \ldots y_n^{s_n}, \qquad 0 \leqq l_i < d_i, \qquad s_i \geqq 0, \qquad i = 1, 2, \ldots, n,$$

where $d_i = \deg f_i$ such that $y_i = f_i(e_i)$ and $R = k[y_1, y_2, \ldots, y_n]$. (In the case (1) y_i and e_i simply coincide.) But $R/R \cap I$ is finite dimensional and hence there exist nonzero polynomials g_i of degree m_i such that $g_i(y_i) = \zeta_i \in R \cap I$, $i = 1, 2, \ldots, n$. Moreover by the preceding paragraph for some positive r_i, $\zeta_i^{r_i} B = \{0\}$. Summing all this up we see that B is spanned over k by a finite number of elements of the form

$$e_1^{l_1} e_2^{l_2} \ldots e_n^{l_n} y_1^{s_1} y_2^{s_2} \ldots y_n^{s_n} \zeta_1^{t_1} \zeta_2^{t_2} \ldots \zeta_n^{t_n} a, \quad 0 \leqq l_i < d_i, \quad 0 \leqq s_i < m_i, \quad 0 \leqq t_i < r_i.$$

Hence B is finite dimensional and the proof is complete.

4.9. Finitely generated algebraic Lie algebras

Let k be a ring with unity and A be a Lie algebra over k. Assume that for each $a \in A$ there exists a polynomial $f_a(t)$ with senior nonzero coefficient 1 such that $f_a(\mathrm{ad}(a)) = 0$. Then A is called *algebraic*, any finite dimensional algebra over some field being evidently algebraic.

In this section we prove two theorems (due to Bahturin-Ol'šanskii) on algebraic Lie algebras having their origin in the group theory.

Theorem 9. *Let k be a field of prime characteristic $p > 0$, A be a finitely generated Lie algebra over k, B be its subalgebra of finite codimension n, $\dim(A/B) = n$. Let, further, there exist $e_1, e_2, \ldots, e_n \in A$ linearly independent modulo B and nonzero $f_1, f_2, \ldots, f_n \in k[t]$ with $f_i(\mathrm{ad}(e_i)) = 0$, $i = 1, 2, \ldots, n$. Then B is a finitely generated Lie algebra.*

Proof. Let first k be algebraically closed and denote by $U = U(A)$ the universal envelope of A. Just as in the proof of the theorem 7.7, one may improve the polynomials f_i so that $z_i = f_i(e_i)$ were central in U and that a basis of U were formed by the elements of the form

$$e_1^{l_1} e_2^{l_2} \ldots e_n^{l_n} z_1^{s_1} z_2^{s_2} \ldots z_n^{s_n} b, \qquad 0 \leqq l_i < \deg(f_i), \qquad s_i \geqq 0, \qquad b \in U(B). \qquad (1)$$

Let I be the linear subspace of U spanned by the elements of the form (1) with $s_1 + s_2 + \dots + s_n > 0$. Then I is easily seen to be an ideal of U such that $U(B) \cap I = A \cap I = \{0\}$, hence A and $U(B)$ can be isomorphically embedded in $Q = U(A)/I$.

Consider now a Lie p-subalgebra G (respectively H) of a Lie p-algebra Q generated by A (respectively by B). Then, obviously, G is a finitely generated Lie p-algebra and, moreover, $\dim(G/H) < \infty$. We prove this latter assertion. Indeed, it suffices to show that the system

$$e_i^{p^{s_i}}, \qquad i = 1, 2, \dots, m, \qquad 0 \leqq s_i < \deg f_i,$$

spans G modulo H over k. However, by the proposition 1,14.2 G is spanned over k by the p^s-powers of the elements $e_1, e_2, \dots, e_n, b \in B$. Since $f_i(e_i) = 0$, $i = 1, 2, \dots, m$, and $b^p \in H$ by the definition of H, our assertion follows. Now using the last corollary from section 2,5, one observes that H is a finitely generated Lie p-algebra.

Note by the way that by the already quoted proposition any element of H is a linear combination of p^s-powers of elements from B, so that we may assume H to possess a finite system of p-generators $b_1, b_2, \dots, b_r$, $b_i \in B$. Prove that the same elements generate B as a usual Lie algebra. Let then b be some element from B and assume that

$$b = \sum_i \alpha_{i0} w_{i0} + \sum_i \alpha_{i1} w_{i1}^p + \dots + \sum_i \alpha_{it} w_{it}^{p^t}, \tag{2}$$

w_{ij} being the values of basic monomials of some rank r free Lie algebra L in $b_1, b_2, \dots, b_r$ (cf. proposition 2,5.1). Since k is algebraically closed, there exist $\beta_{ij} \in k$ such that $\alpha_{ij} = \beta_{ij}^p$ for all indices i, j. But then by proposition 1,14.1, there exists an element b' from the ordinary Lie subalgebra of H generated by $b_1, b_2, \dots, b_r$, that is $b' \in B$, such that

$$b' + \sum_i \alpha_{i1} w_{i1}^p + \dots + \sum_i \alpha_{it} w_{it}^{p^t} = \Big(\sum_i \beta_{i1} w_{i1} + \dots + \sum_i \beta_{i1} w_{it}^{p^{t-1}}\Big)^p. \tag{3}$$

By (2) the left hand side of (3) lies in B and the left hand side is a p-power of some element from $U(B)$. But using the corollary from the PBW-theorem 2, 4.6, one readily understands that non nonzero element from B is representable as a p-power ($p > 1$) of an element from $U(B)$. Thus the left hand side of (3) is zero and $b = \sum \alpha_{i0} w_{i0} - b' \in B$ so that in the case of algebraically closed fields the proof is complete.

Assume, second, that k is (possibly) not algebraically closed and k^* is its algebraic closure. Embed A canonically as a k-subspace into $A_{k^*} = k^* \otimes_k A$ and denote by B_{k^*} the subalgebra of A_{k^*} generated by B and equal to $k^* \otimes_k B$. It is a routine calculation to check up that A_{k^*} and B_{k^*} satisfy the conditions of the theorem so that by the first part of the proof there exist $b_1^*, b_2^*, \dots$ b_s^* generating B_{k^*} as a Lie subalgebra over k^*. Each of these elements however has the form

$$b_i^* = \alpha_{i1} \otimes b_{i1} + \dots + \alpha_{is_i} \otimes b_{is_i}, \quad i = 1, 2, \dots, s, \qquad \alpha_{ij} \in k^*, \quad b_{ij} \in B.$$

Now it is clear that B can be generated by a system of elements $\{b_{ij}, i = 1, 2, \dots, s; j = 1, 2, \dots, s_i\}$, and the proof of the theorem is complete.

Remark that without an 'algebraicity condition' the conclusion of the theorem (even when A is metabelian) is wrong. For example, let $A = L(\{x, y\}, \mathfrak{A}^2)$, the free metabelian algebra. Then $\dim(A/A^2) = 2$, while A^2 is a free abelian algebra of infinite rank (see 3,6.6).

An important corollary is the next

Theorem 9'. *Let k be a finite commutative ring with unity and A a finitely generated algebraic Lie algebra over k (see the definition above). Given a subalgebra B of finite index as an abelian subgroup of A, B is a finitely generated Lie algebra.*

Proof. Let J be the Jacobson's radical of k, that is the maximal nilpotent ideal of k. Using the same argument as in the proposition 1,8.1 one deduces that B can be generated by any set of elements generating B modulo JB. Moreover, since $B \cap JA/JB$ is a finite module, it suffices to prove that $B/B \cap JA$ is a finitely generated algebra. Since, further, $B/B \cap JA \cong B + JA/JA = \overline{B} \subseteq \overline{A} = A/JA$, we can limit ourselves to the case of semisimple k (with $J = \{0\}$). Let then $\varepsilon_1, \varepsilon_2, \dots, \varepsilon_n$ be a complete system of pairwise orthogonal idempotents of k, that is $k = k\varepsilon_1 \oplus k\varepsilon_2 \oplus \dots \oplus k\varepsilon_n$, and $k\varepsilon_i$, is a field, $i = 1, 2, \dots, n$. Clearly, $A = \sum_{i=1}^{n} \varepsilon_i A \supset \sum_{i=1}^{n} \varepsilon_i B = B$, so that each $\varepsilon_i B$ is a subalgebra of finite index in $\varepsilon_i A$, this latter being already an algebra over a field $k\varepsilon_i$. Therefore theorem 9 applies and hence we are finished.

The final theorem of this chapter needs more restrictive conditions.

Theorem 10. *Let k be a finite commutative ring with unity, A be some finitely generated Lie algebra over k, f be a polynomial with senior coefficient 1 from $k[t]$ such that $f(\mathrm{ad}(a)) = 0$ for any $a \in A$. Then for each positive integer n there exist only finitely many subalgebras having index n in A.*

Proof. We show that, if k is a field, a 'p-polynomial' can be found to annihilate all the $\mathrm{ad}(a)$, $a \in A$. For, if f has degree m, then $t^{p^i} = f(t)\, q_i(t) + r_i(t)$, $i = 1, 2, \dots, m$, $\deg(r_i) < m$. Since the subspace of all polynomials of degree $< m$ is m dimensional, there exist $\alpha_1, \alpha_2, \dots, \alpha_m \in k$ not all zero such that $\sum_{i=0}^{m} \alpha_i r_i = 0$. But then $f(t)$ is a divisor of a p-polynomial $\sum_{i=1}^{m} \alpha_i t^{p^i}$. Thus we assume that in the case of a field $f(t)$ itself is a p-polynomial and, moreover, that $f(t)$ has no constant term and is of degree $d \geqq p$.

To prove the theorem in the case when k is a finite field we correspond to each subalgebra B of index n a linear representation of A having degree $(n + 1)\, d^n$ and show that for different B these representations are not equivalent. Since A is finitely generated and $k_{(n+1)d^n}$ is finite this will imply the conclusion of the theorem.

To this end, fix B and choose a basis $e_1, e_2, \dots, e_n$ complementing some basis of B. Denoting by z_i the elements $f(e_i)$ we get as before that the basis of $U(A)$ is formed by

$$e_1^{s_1} e_2^{s_2} \dots e_n^{s_n} z_1^{l_1} z_2^{l_2} \dots z_n^{l_n} b \,, \tag{4}$$

where $0 \leqq s_i < d$, $l_i \geqq 0$, $i = 1, 2, \dots, n$, b is some basic element of $U(B)$. Let V_B be the subspace of $U = U(A)$ spanned by those monomials (4) in which either $b \neq 1$ or $l_1 + l_2 + \dots + l_n \geqq 2$. Then V_B is easily seen to be an A-submodule of a left A-module U. Denote by ϱ_B the corresponding representation of A in U/V_B. This is the desired representation.

Indeed, we show that $\varrho_B(f(x)) = 0$ if $x \in B$ and $\varrho_B(f(x)) \neq 0$ if $x \notin B$. The former statement is evident since $f(x)$ is central in U and is an element of $U(B)$ at the same time. To prove the latter one considers the filtration $(U_m)_{m \geqq 0}$ of U introduced in 2,4. Then, given an element $x = \alpha_1 e_1 + \alpha_2 e_2 + \dots + \alpha_n e_n + h$, $h \in B$, $\alpha_i \in k$, $i = 1, 2, \dots, n$, and using the fact that f is a p-polynomial, $p = \mathrm{char}(k)$, one has

$$f(x) = f(\alpha_1 e_1) + f(\alpha_2 e_2) + \dots + f(\alpha_n e_n) + f(h) + u \,, \qquad u \in U_{d-1} \,. \tag{5}$$

Remembering that $f(e_i) = z_i$, $i = 1, 2, \dots, n$, one rewrites (5) in the form

$$f(x) = a_1^t z_1 + \alpha_2^t z_2 + \dots + \alpha_n^t z_n + f(h) + v \,, \qquad v \in U_{d-1} \,. \tag{6}$$

The latter expression (6) shows that

$$\varrho_B(f(x))\,(1 + V_B) = f(x) + V_B \neq 0$$

if for some i, $\alpha_i \neq 0$. Thus the case when k is a field is accomplished.

Let further k be a finite commutative ring with Jacobson's radical J, $J^l = \{0\}$. If $J = \{0\}$ and $\varepsilon_1, \varepsilon_2, \dots, \varepsilon_s$ is a complete system of pairwise orthogonal idempotents, $A = \varepsilon_1 A \oplus \varepsilon_2 A \oplus \dots \oplus \varepsilon_s A$, where $\varepsilon_i A$ is a finitely generated Lie algebra over $\varepsilon_i k$. Since in our case any subalgebra B is the direct sum of its projections onto $\varepsilon_1 A, \varepsilon_2 A, \dots$ $\dots, \varepsilon_s A$, the case $J = \{0\}$ is implied by the preceding argument valid for fields. If $J \neq \{0\}$ then from the just proved one deduces that there are only finitely many subalgebras of the form $B + JA$, B being any subalgebra of index n in A. Taking this assertion for the base of induction, we prove that there is only a finite number of subalgebras of the form $B + J^s A$, $|A:B| = n$. Indeed, let $C = B + J^s A$ be one of these subalgebras. Then $C = B + JH$, where $H = B + J^{s-1}A$. But being a subalgebra of finite index in A, H is a finitely generated algebra (theorem 9′), hence $|H/B| \leqq n$ implies the finiteness of the number of subalgebras having the form C in H. But by the induction the number of subalgebras of the form H is itself finite, which obviously proves the inductive assertion.

Taking now $s = l$ we get the conclusion of the theorem.

The reader is suggested to construct easy examples showing that the restrictions in this theorem are indispensable.

CHAPTER 5

Identities in finite Lie algebras

5.1. A characterization of a variety generated by a finite Lie algebra

Let k be a finite commutative ring and A be a finite Lie algebra over k. Denote by $\mathfrak{V} = \text{var}(A)$ the variety generated by A, that is the variety determined by identical relations holding in A. By the corollary from 3,1 $\mathfrak{V}$ is locally finite. In this section we give some more precise characterisation of $\mathfrak{V}$.

To do this we introduce a class $\mathfrak{C}(f, m, c)$ of Lie algebras over k determined by two positive integers m, c, and a polynomial f in one variable with major coefficient 1.

An algebra L over k belongs to $\mathfrak{C}(f, m, c)$ if

(i) for any $x \in L$, $f(\text{ad}(x)) = 0$;

(ii) for any chief factor M of L, $|M| \leqq m$;

(iii) for any nilpotent factor N of L, $N^{c+1} = \{0\}$.

We shall see in the future that each such class is a finitely based variety generated by a finite Lie algebra. But now we can prove in this section only

Proposition 1. *Let A be a finite algebra from some class $\mathfrak{C}(f, m, c)$, $\mathfrak{V} = \text{var}(A)$. Then $\mathfrak{V} \subseteq \mathfrak{C}(f, m, c)$.*

Proof. Assuming $L \in \mathfrak{V}$ we show that L satisfies (i)–(iii).

(i) is trivial, since this condition is equivalent to the identical relation $(f(\text{ad}(x))\,(y) \equiv 0$ holding in A and hence in any algebra from $\mathfrak{V}$.

(ii) Suppose first that L is finite. By a construction of Birkhoff's theorem 3,1.1, $L \cong B/C$, $B \subseteq P = \prod_{i=1}^{n} A_i$, $A_i \cong A$, $i = 1, 2, \dots, n$, n be some finite positive integer. Any chief factor of P is isomorphic (as an abelian group) with a chief factor of $A_i \cong A$ and therefore is bounded by m. Since by the corollary in 1,15 the chief factors of a subalgeb ra and a quotient algebra are isomorphic (as abelian groups) to subfactors of chief factors of the whole algebra the condition (ii) in this case holds.

Let second L be infinite and assume that M is a minimal ideal of L of order $> m$. Consider then a finite k-submodule $N \subseteq M$ with $|N| > m$. For any $0 \neq x \in N$ there exists a finite number of elements $a_1, a_2, \dots, a_s$ from L such that $\text{alg}(\{a_1, a_2, \dots \dots, a_s, x\}) \supseteq N$. Let then D be a finitely generated (and thus finite) subalgebra of L containing N such that any ideal of D having nonzero intersection with N contains this latter. Then as soon as T is a maximal ideal of D with $N \cap T = \{0\}$ and S is a minimal ideal of D containing T, the chief factor S/T has order $\geqq |N| > m$ which contradicts the above considered situation.

(iii) Since any factor of an algebra from $\mathfrak{V}$ is itself an algebra from $\mathfrak{V}$ and as an algebra has nilpotency class c if and only if all its $(c + 1)$-generator subalgebras are such, we may limit ourselves to the case of a nilpotent $(c+1)$-generator (and thus finite) algebra L, representable, as in (ii), in the form

$$L \cong B/C\,, \qquad C \subseteq P = \prod_{i=1}^{n} A_i\,, \qquad A_i \cong A\,, \qquad i = 1, 2, \ldots, n\,. \tag{1}$$

Let firstly $C \subseteq \Phi(B)$, the Frattini subalgebra of B. Then by theorem 1,10.5 B is nilpotent and so are its projections onto A_i, having by our assumption about A, classes $\leqq c$. Now B and consequently L have class $\leqq c$ as required.

Suppose secondly, $C \nsubseteq \Phi(B)$ and let B be minimal such that (1) takes place. By our assumption for some maximal subalgebra M of B one has $B = M + C$. This however implies $L \cong (M + C)/C \cong M/M \cap C$ which contradicts the minimality of B. Therefore (iii) holds and the proof of the proposition is complete.

One more piece of knowledge about $\mathfrak{C}(f, m, c)$ gives proposition 2 which we anticipate by two simple lemmas.

Lemma 1. *Let M be a chief factor of some Lie algebra L such that $|M| \leqq m$. Then $|L/C_L(M)| \leqq m^m$.*

Proof. This is a specification of proposition 4,4.1(iii).

Lemma 2. *Let G be a finitely generated nonzero ideal of L. Then there exists a maximal ideal H of L contained in G with $H \neq G$.*

Proof. If not, G is a union of some infinite properly increasing chain of L-ideals H_α, $\alpha = 1, 2, \ldots$. But being finitely generated as an ideal, G then is contained in H_α containing all these generators, which is impossible.

Proposition 2. *Class $\mathfrak{C}(f, m, c)$ is locally finite.*

Proof. Choose a finitely generated Lie algebra $L \in \mathfrak{C}$ and denote by F the intersection of all chief centralizers of L, that is centralizers of chief factors of L. By lemma 1 and theorem 4,9.10 there exist only finitely many such centralizers, so that $|L/F| < \infty$. Using now theorem 4,9.9 one sees that F is an ideal of L finitely generated as subalgebra. We denote also by G the intersection of all ideals of finite index in L and show that (i) $|L/G| < \infty$, (ii) $G = \{0\}$. So far let N be some ideal of finite index in L and $N_1 = F \cap N$. Consider a chief series of L/N_1 which is a refinement of a normal series $L/N_1 \rhd F/N_1 \rhd N_1/N_1$. By the definition of F, all the chief factors which lie between F/N_1 and N_1/N_1 are central and thus F/N_1 is nilpotent of class $\leqq c$. Since k is finite, the order of F/N_1 is bounded only by $|k|$, c and the number of generators in F. Again, by theorem 4,9.10, there exist only finitely many subalgebras N_1 in F and hence $|L/N|$ is also a bounded value. This proves (i). The second statement now follows immediately. For, by theorem 4,9.9, G is finitely generated and if $G \neq \{0\}$, there exists an $H \lhd L$ such that G/H is a chief factor of L. By the definition of G, $|L/H|$ and hence $|G/H|$ are infinite, thus contradicting the definition of $\mathfrak{C}$.

So $L = L/G$ is finite and the proof is complete.

5.2. Critical algebras

Given a finite Lie algebra A over k denote by $(\mathrm{Qs} - 1)\,A$ the set of all its proper factors. An algebra A is called *critical* if $A \notin \mathrm{var}(\mathrm{Qs} - 1)\,A$. By proposition 4,8.1 (i) a critical algebra is necessarily monolithic, the converse being wrong. For a counterexample of a monolithic but not critical algebra one may take an algebra from the example in 4,5. The use of critical algebras may be explained by

Proposition 1. *Any locally finite variety is generated by its critical algebras.*

Proof. Since any free algebra of any variety $\mathfrak{V}$ is residually $\mathfrak{V}$-free of finite rank and by the local finiteness of $\mathfrak{V}$, this latter is generated by its finite algebras forming a set, say, M. Consider then the set $M' = \mathrm{QS}(M)$ of all factors of algebras from M. Since $\mathrm{QS}(M') = \mathrm{QS}(M) = M'$, any noncritical algebra $A \in M'$ belongs to $\mathrm{var}(M' \setminus \{A\})$. Finally, $\mathfrak{V}$ is generated by the subset of critical algebras from M', and this finishes the proof.

An important criterion of non-criticity of algebras gives

Proposition 2. *Let A be a finite Lie algebra possessing a set of ideals $M_1, M_2, \dots, M_s$ and a subalgebra L such that*

(i) $A = \mathrm{alg}(L, M_1, M_2, \dots, M_s)$;

(ii) $A \neq \mathrm{alg}(L, M_1, \dots, \hat{M}_i, \dots, M_s)$, *where means that this ideal is omitted, for any $i = 1, 2, \dots, s$;*

(iii) $[M_{\pi(1)}, M_{\pi(2)}, \dots, M_{\pi(s)}] = \{0\}$ *for any substitution π of indices* $1, 2, \dots, s$.

Proof. Put $B_0 \cong L$, $B_i \cong M_i$, $i = 1, 2, \dots, s$, and $P = B_0 * B_1 * \dots * B_s$ (see 2, 6.1). Denote by θ_j an endomorphism of P induced by the identity map $B_i \to B_i$, where $i \neq j$, and zero map $B_j \to \{0\}$; let also $K = \bigcap_{j=1}^{s} \mathrm{Ker}\,\theta_j$. Clearly, any element from K is representable as a linear combination of commutators each of them includes factors from each $B_1, B_2, \dots, B_s$. Therefore, if α is a homomorphism of P onto A induced by the isomorphisms $B_0 \to L$, $B_i \to M_i$, $i = 1, 2, \dots, s$, then by (iii) $K \subseteq \mathrm{Ker}\,\alpha$.

Now consider an element $w = (1 - \theta_1)(1 - \theta_2) \dots (1 - \theta_s)(u)$, u being some element of P. Then, clearly, $w \in K$, and thus we have

$$u = w + v_1 + v_2 + \dots + v_t, \tag{1}$$

where $v_i = \pm\beta_i(u)$, β_i be some *nonempty* product of endomorphisms θ_i, $i = 1, 2, \dots, t$; $w \in \mathrm{Ker}\,\alpha$. Denote now $A_i = \alpha\beta_i(P)$. Then the condition (ii) shows that A_i, $i = 1, 2, \dots, t$, is a proper subalgebra of A. To complete the proof of the proposition it suffices to prove that A is representable as a homomorphic image of a direct product $S = A_1 \times A_2 \times \dots \times A_t$. So far, represent S as a quotient algebra of P by putting $\varphi(u) = (\alpha(v_1), \alpha(v_2), \dots, \alpha(v_t))$ for any $u \in P$ of the form (1). Evidently from the definitions $\mathrm{Ker}\,\varphi \subseteq \mathrm{Ker}\,\alpha$ and thus theorem 1,2.1 (iii) applies showing that A is a quotient algebra of S. This completes our proof.

The final assertion of this section is an auxiliary

Lemma. *Let $M_1, M_2, \dots, M_s$ be nonnilpotent ideals of a finite Lie algebra A such that $\mathrm{alg}(M_1, M_2, \dots, M_s) = M_1 \times \dots \times M_s$. Then A possesses a subalgebra L with $A = \mathrm{alg}(L, M_1, M_2, \dots, M_s)$ and $A \neq \mathrm{alg}(L, M_1, \dots, \hat{M}_i, \dots, M_s)$, $i = 1, 2, \dots, s$.*

Proof. Let L be a minimal subalgebra of A with $A = \mathrm{alg}(L, M_1, \dots, M_s)$ and without loss of generality assume that also $A = \mathrm{alg}(L, M_2, \dots, M_s)$. Denote $M = M_1 \times M_2 \times \dots \times M_s$ and $M^* = M_2 \times \dots \times M_s$, then $A = L + M = L + M^*$. By the modular law, $M = M^* + M \cap L$. If now $N = M^* \cap L$, $M_1 \cong$

$\cong M/M^* = M \cap L + M^*/M^* \cong M \cap L/M^* \cap L = M \cap L/N$, and therefore the ideal $M \cap L/N$ is not nilpotent in L/N. By theorem 1,10.5, $M \cap L/N \nsubseteq \Phi(L/N)$ so that for some maximal proper subalgebra L_1/N of L/N one has $M \cap L/N + L_1/N = L/N$. This latter equality yields $L = L_1 + M \cap L$ and further $A = L + M = L_1 + M \cap L + M = L_1 + M$ which contradicts the minimality of L. Thus the proof of the lemma is complete.

5.3. Minimal representation

A set $\mathcal{S}$ of algebras is called a *factor closed* one if $\mathrm{QS}(\mathcal{S}) = \mathcal{S}$. Let $\mathfrak{U} = \mathrm{var}(\mathcal{S})$ where $\mathcal{S}$ is a finite factor closed set of finite Lie algebras over k. Then any finite algebra $A \in \mathfrak{U}$ by the construction of the theorem 3,1.1 is isomorphic to a factor of a finite direct product P of algebras from $\mathcal{S}$:

$$A = B/C\,, \qquad B \subseteq P = \prod_{i=1}^{n} S_i\,, \qquad S_i \in \mathcal{S}\,, \qquad i = 1, 2, \ldots, n\,. \tag{1}$$

Assume without loss of generality that $|S_1| \geqq |S_2| \geqq \ldots \geqq |S_n|$. Then the representation of A in the form (1) is called *minimal* if the string of orders $(|S_1|, |S_2|, \ldots, |S_n|)$ is lexicographically the least possible.

Some properties of a minimal representation are given in the following

Proposition 1. *Let* (1) *be a minimal representation of a finite algebra A in a variety $\mathfrak{U}$ generated by a finite factor closed set $\mathcal{S}$ of finite algebras. Then*

(i) *All algebras S_i, $i = 1, 2, \ldots, n$, are critical;*
(ii) *B is a subdirect product in P;*
(iii) *A subalgebra $D \subseteq S_i$ is an ideal of S_i if and only if $[B, D] \subseteq D$;*
(iv) *Any nontrivial ideal $D \subseteq S_i$ has nontrivial intersection with B;*
(v) *$C \cap S_i = \{0\}$ for all $i = 1, 2, \ldots, n$.*

Proof. (i) If some S_i is not critical then S_i is isomorphic to a factor of a finite direct product of its proper factors having smaller orders. This leads to a contradiction with the minimality of (1).

(ii) If π_i is a projection of P onto S_i, $i = 1, 2, \ldots, n$, then B is in fact a subalgebra of $\pi_1(B) \times \pi_2(B) \times \ldots \times \pi_n(B)$. The minimality of (1) implies therefore $\pi_i(B) = S_i$, $i = 1, 2, \ldots$.

(iii) follows immediately from (ii).

(iv) If not, then for some nontrivial ideal D of, say S_1, one has $D \cap B = \{0\}$ and B is isomorphic to a subalgebra of $S_1/D \times S_2 \times \ldots \times S_n$, contradicting the minimality of (1).

(v) Assume $\{0\} \neq D \lhd S_1$ is such that $D \cap C \neq \{0\}$. Then A is isomorphic to a factor of $S_1/D \cap C \times S_2 \times \ldots \times S_n$.

This finishes the proof of the proposition.

We recall one more definition from 1,15. An invariant series

$$0 = A_0 \subseteq A_1 \subseteq \ldots \subseteq A_q = A \tag{2}$$

of a finite Lie algebra A is called a *fully reducible* one if each of its factors A_{i+1}/A_i, $i = 0, 1, \ldots, q-1$, is a direct product of chief factors of A. The number q is called the length of (2). An important application of the concept of minimal representation is

Proposition 2. *Let A be a finite Lie algebra from the variety $\mathfrak{U}$ generated by a finite factor closed set $\mathcal{S}$ of finite algebras such that the orders of critical algebras from $\mathcal{S}$ are bounded by some positive integer q. Then A possesses a fully reducible series of a length $\leqq q$.*

Proof. Let (1) be a minimal representation of A in $\mathfrak{U}$. Then by 1(i) all S_i are critical algebras with monoliths M_i. By 1(iii) and 1(iv), each M_i is a minimal ideal in B, $i = 1, 2, \ldots, n$. Denote by Q_1 the direct product $Q_1 = M_1 \times M_2 \times \ldots \times M_n$; clearly B/Q_1 is isomorphic to a subalgebra of $S_1/M_1 \times S_2/M_2 \times \ldots \times S_n/M_n$ and one may use induction by q. By the induction assumption B/Q_1 possesses a fully reducible series of a length $\leqq q - 1$ the inverse images of whose members together with Q_1 form a fully reducible series

$$\{0\} = Q_0 \subseteq Q_1 \subseteq \ldots \subseteq Q_q = B$$

of B. To obtain a fully reducible series of B/C it suffices to apply the corollary of 1,15.

5.4. Critical algebras in a class $\mathfrak{C}(f, m, c)$

We use in this section the structure theory of finite Lie algebras developed in 1,13. Thus let A be a finite Lie algebra, N be its nilradical, φ be its Frattini ideal, $M_1, M_2, \ldots, M_s$ be ideals of A such that $N/\varphi = M_1/\varphi \oplus M_2/\varphi \oplus \ldots \oplus M_s/\varphi$, M_i/φ be a minimal abelian ideal of A/φ, L be a subalgebra of A such that $A/\varphi = M/\varphi \oplus L/\varphi$ (see 1,13.9). Let also C be a centralizer of N/φ in A and F be the intersection of all chief centralizers.

Lemma 1. *$F = N$. Moreover, if A is solvable, $C = N$.*

Proof. Since F centralizes its A-chief series, it possesses a central series and thus is nilpotent. Hence $F \subseteq N$. Consider now the lower central series of N; its terms are A-invariant and therefore one may consider a chief series of A refining the A-series $A \rhd N \rhd N^2 \rhd \ldots \rhd N^{d+1} = \{0\}$. Any term of this series is situated either 'over' N or 'inside' N and in both cases is centralized by N. But any chief factor of a finite algebra by the corollary of 1,15. is isomorphic to some factor of any fixed chief series and hence is centralized by N. Finally $N \subseteq F$ and the proof is complete in the general case.

Assume then A to be solvable. If $C \neq N$ then one can consider an abelian chief factor D/N contained in C/N. Since D centralizes N/φ and N/φ is nilpotent, so is D/φ. In this case, however, by theorem 1, 10.5 D is itself nilpotent, contradicting the definition of N as a largest nilpotent ideal of A. The lemma is thus proved.

Lemma 2. *Let A be a critical algebra from a class $\mathfrak{C}(f, m, c)$. Then $|N/\varphi| \leqq m^c$ and $|A/C| \leqq m^{mc}$. If A is solvable then its order is bounded by a number depending only on $\mathfrak{C}$. So $\mathfrak{C}$ possesses only finitely many solvable critical algebras.*

Proof. If $|N/\varphi| > m^c$, then proposition 2.1 applies to the system $M_1, M_2, \ldots, M_s, L$, for obviously $|M_i/\varphi| < m$ and therefore $s > c$, whence $[M_{\pi(1)}, M_{\pi(2)}, \ldots, M_{\pi(s)}] = \{0\}$. Two other conditions of this proposition are evident. So $|N/\varphi| \leqq m^c$ and $s \leqq c$. But C is an intersection of centralizers of the chief factors M_i/φ, $i = 1, 2, \ldots, s$. Using lemma 1.1, $|A/C| \leqq (m^m)^c$.

If A is solvable, by lemma 1, $|A/\varphi| \leqq m^{(m+1)c}$, and thus A may be generated by $\leqq m^{(m+1)c}$ elements (see 1,8). Consider then a free algebra of the variety determined by $(f(\mathrm{ad}(x)))\,(y) \equiv 0$ of rank $\leqq m^{(m+1)c}$. By theorem 4,9.9 its subalgebra of index $\leqq m^{(m+1)c}$ has at most $a = a(f, m, c)$ generators. The same value a bounds, clearly, the number of generators of φ. Since φ is nilpotent of class $\leqq c$ and the base ring k is finite, the order of φ is bounded by an integer $b = b(f, m, c)$ depending only on $\mathfrak{C}$. So $|A| = |A/\varphi|\,|\varphi|$ is bounded and the proof is complete.

Now we formulate the main result of this section.

Theorem 1. *Any class $\mathfrak{C}(f, m, c)$ possesses only finitely many pairwise nonisomorphic critical algebras.*

Proof. In view of the preceding lemma and its proof, it suffices to show that if A is a critical algebra in $\mathfrak{C}$, N is its nilradical, C is the centralizer of N/φ then $|C/N|$ is bounded by a number depending only on $\mathfrak{C}$.

By lemma 1, N is representable as $N = \bigcap_{i=1}^{t} (C \cap C_i)$, C_i being centralizers of some chief factors of A. Assume that t is minimal in this representation of N; then for each $i = 1, 2, \dots, t$, $H_i = \bigcap_{j \neq i} (C \cap C_j)$ properly contains N and

$$\mathrm{alg}(H_1/N, H_2/N, \dots, H_t/N) = H_1/N \times H_2/N \times \dots \times H_t/N\,.$$

Choose minimal ideals N_i/N of A/N in H_i/N, $i = 1, 2, \dots, t$. By the same reason as in the proof of lemma 1, each N_i/N is nonnilpotent. By the lemma from section 2, there exists a subalgebra L/N which generates A/N with a set $N_1/N, \dots, N_t/N$ but with none of its proper subsets. To apply proposition 2.2 it suffices to show that for some t depending only on $\mathfrak{C}$,

$$[N_{\pi(1)}, N_{\pi(2)}, \dots, N_{\pi(t)}] = \{0\}\,.$$

So far choose $b_1, b_2, \dots, b_t$ from $N_1, N_2, \dots, N_t$ respectively and denote by B a subalgebra generated by these elements and N. B is a solvable algebra and if $S_1, S_2, \dots, S_l$ are its critical factors then their orders by lemma 2 are bounded by a number q depending only on $\mathfrak{C}$. Applying proposition 3.2 to $B \in \mathrm{var}(\mathrm{QS}(B))$, B possesses a fully reducible series of a length $\leqq q$. Clearly, a fully reducible B-series of a length $\leqq q$ may be chosen in the B-ideal N

$$\{0\} = Z_0 \trianglelefteq Z_1 \trianglelefteq \dots \trianglelefteq Z_q = N\,. \tag{1}$$

Assuming now $t = 1 + qm^m$, choose a subset, say, $b_1, b_2, \dots, b_{m_m}$ of $b_1, b_2, \dots, b_t$ and a chief B-factor T_λ/Z_r from a B-series (1).

We prove that given $x \in T_\lambda$,

$$[b_1, b_2, \dots, b_{m^m}, x] \in Z_r\,. \tag{2}$$

Indeed, any B-factor is B-isomorphic to a B-subfactor of a chief A-factor (see 1, 15, corollary). Thus T_λ/Z_r is B-isomorphic to a B-subfactor of some chief A-factor X_λ/Y_λ. Let C_λ be the centralizer of X_λ/Y_λ and $U = (C_\lambda/N) \cap (N_1/N \times \dots \times N_{m_m}/N)$. We show that U is a product of some of N_i/N, $1 \leqq i \leqq m^m$. For if $0 \neq u + N \in U$, then $u = \sum_i d_i + N$, $d_i \in N_i$; if, further, $d_1 \notin N$, then as N_1/N is minimal and nonabelian, for some $d_1' \in N_1$, $0 \neq [u + N, d_1' + N] = [d, d_1'] + N \in N_1/N$. By the minimality of N_1/N, this latter now evidently lies in U. Since $|A/C_\lambda| \leqq m^m$, the same is true for $(N_1/N \times$

$\times \ldots \times N_{mm}/N))/U$ and, by the just proved, some of $b_1, b_2, \ldots, b_{mm}$, say, b_j belongs to C_λ. Thus $[b_j, T_\lambda] \subseteq Z_r$, and since T_λ, Z_r are B-invariant, (2) follows.

By the definition of (1) each factor in this series is a direct sum of chief factors of the above considered form T_λ/Z_r. Thus the preceding argument gives $[b_1, b_2, \ldots, b_{mm}, Z_r] \subseteq Z_{r+1}$. Since B/N is abelian this implies

$$[b_1, b_2, \ldots, b_{qmm}, b] = [b_1, b_2, \ldots, b_t] = 0 .$$

This enables us to apply proposition 3.2 if $t > 1 + qm^m$. So $t \leq 1 + qm^m$ and thus $|C/N|$ is bounded only by $\mathfrak{C}$.

Now the proof of the theorem is complete.

An important consequence of this theorem is

Theorem 2. *$\mathfrak{C}(f, m, c)$ is a variety.*

Proof. Denote by A the finite direct product of all critical algebras from $\mathfrak{C}$ and by $\mathfrak{V}$ the variety var(A) generated by A. To prove $\mathfrak{V} = \mathfrak{C}$ it suffices to show that any finitely generated subalgebra B of any algebra $C \in \mathfrak{C}$ belongs to $\mathfrak{V}$. If C is finitely generated then C is finite by proposition 1,2, and B lies in the variety generated by its critical factors which by proposition 1.1 are critical algebras of $\mathfrak{C}$. Clearly then $B \in$ var$(A) = \mathfrak{V}$.

Suppose now C be an arbitrary algebra from $\mathfrak{C}$ and $V(B)$ be nonzero. Again by proposition 1.1, $B/V(B) \in$ var$(A) \subseteq \mathfrak{C}$. In particular, $B/V(B)$ is finite and $V(B)$ is finitel ygenerated as an ideal of B (see for example the proof of theorem 4,8.8). Now similarly to the proof of the proposition 1.2, there exists a maximal ideal N of C (and hence of B) such that $N \not\supseteq V(B)$. Let M be the intersection of all C-ideals properly containing N; by the choice of N, $M \supseteq V(B)$ and, moreover, M/N is a chief factor of C. Thus, $|M/N| \leq m$. Put now $L = V(B) \cap N$ and consider B/L. Since $V(B)/L \cong V(B) + N/N \subseteq M/N$ and as any chief factor of B/L is isomorphic either to a chief factor of B/V or to a B-factor of V/L, all chief factors of B/L have order $\leq m$. Further, being a factor of C, B/L is bounded by parameters c and f. Hence $B/L \in \mathfrak{C}$ and by the first part of the proof $B/L \in \mathfrak{V}$. On the other hand $V(B/L) = V(B)/L \neq \{0\}$, which is a contradiction proving the theorem.

5.5. Varieties $\mathfrak{V}^{(n)}$

Let $\mathfrak{V}$ be a variety and n be a positive integer. Let V_n be the set of all identities holding in $\mathfrak{V}$ and depending on at most n variables. Then by $\mathfrak{V}^{(n)}$ we mean the variety of all algebras satisfying all identities $v \equiv 0$, $v \in V_n$. Equivalently, $\mathfrak{V}^{(n)}$ is the class of all algebras whose r-generator subalgebras are in $\mathfrak{V}$, $0 \leq r \leq n$. Clearly, $\mathfrak{V} \subseteq \mathfrak{V}^{(n)}$ for all $n = 1, 2, \ldots$.

If $\mathfrak{V}$ is a finitely based variety (see the final paragraph of 3,2) then $\mathfrak{V}$ may be determined by a single identity depending on, say, n variables. Then $\mathfrak{V} = \mathfrak{V}^{(n)}$. A partial converse of this remark is the following proposition, explaining together with the proposition below at the same time the significance of these varieties for the study of identities in finite algebras.

Proposition 1. *Let $\mathfrak{V}$ be a locally finite variety over some ring k and n be a positive integer. Then $\mathfrak{V}^{(n)}$ is a finitely based variety.*

Proof. Indeed, given a free Lie algebra $L = L(X)$ of countable rank, its subalgebra $L' = L(\{x_1, x_2, \ldots, x_n\})$, $x_i \in X$, and a verbal subalgebra $V = V(L)$ determining $\mathfrak{V}$, one immediately sees that V_n is equivalent to $V \cap L' = V(L')$ (this is explained in the final paragraph of the proof of proposition 3,4.8). By our condition $L'/V(L')$ is finite and hence $V(L')$ is finitely generated as an ideal of L' (see for example the proof of the theorem 4,8.8), and all the more as a verbal subalgebra of L'.

This completes our proof.

Proposition 2. *Assume that $\mathfrak{U}$ is a finitely based subvariety of $\mathfrak{C}(f, m, c)$, and the variety $\mathfrak{W}$ is such that $\mathfrak{W} \subseteq \mathfrak{U}$. Then $\mathfrak{W}$ is itself finitely based.*

Proof. By proposition 2.1 any locally finite variety and in particular, $\mathfrak{C}$, $\mathfrak{W}$ and $\mathfrak{U}$ (see proposition 1.2) are generated by their critical algebras. Theorem 1 proves thus that $\mathfrak{C}(f, m, c)$ has a finite number of subvarieties. In particular, their exists a chain

$$\mathfrak{W} = \mathfrak{W}_0 \subset \mathfrak{W}_1 \subset \ldots \subset \mathfrak{W}_{s-1} \subset \mathfrak{W}_s = \mathfrak{U}$$

of subvarieties such that there exists no subvariety $\mathfrak{V}$ with $\mathfrak{W}_i \subsetneqq \mathfrak{V} \subsetneqq \mathfrak{W}_{i+1}$, $i = 0, 1, 2, \ldots, s-1$. Proceed then by induction over s, the case $s = 0$ being trivial. Let $u_1 \equiv 0, \ldots, u_t \equiv 0$ be a finite basis of identities holding in $\mathfrak{U}$ and $u_{t+1} \equiv 0$ be an identity of $\mathfrak{W}_{s-1}$ which is not satisfied by $\mathfrak{U} = \mathfrak{W}_s$. Let $\mathfrak{V}$ be the variety determined by a system $u_1 \equiv 0$, $u_2 \equiv 0, \ldots, u_t \equiv 0$, $u_{t+1} \equiv 0$. Then $\mathfrak{W}_{s-1} \subseteq \mathfrak{V} \subsetneqq \mathfrak{W}_s$ and hence $\mathfrak{V} = \mathfrak{W}_{s-1}$. Therefore $\mathfrak{W}_{s-1}$ is finitely based and by the induction hypothesis then $\mathfrak{W}$ is ditto.

Corollary. *Let A be a finite algebra and $\mathfrak{V} =$ var (A). Then $\mathfrak{V}$ is finitely based if and only if for some n, f, m, c, $\mathfrak{V}^{(n)} \subseteq \mathfrak{C}(f, m, c)$.*

Proposition 3. *Let A be a finite algebra and $\mathfrak{V} =$ var $(A) \subseteq \mathfrak{C}(f, m, c)$. Then there exists a $s > 0$ such that any finite algebra $G \in \mathfrak{V}^{(s)} \cap \mathfrak{C}^2$ lies in $\mathfrak{C}$.*

Lemma 1. *Assume $n \geqq m + 1$. Then there exists a $q = q(\mathfrak{V})$ such that for any monomial $[x_{i_1}, x_{i_2}, \ldots, x_{i_s}, y]$, $i_j \in \{1, 2, \ldots, m\}$ there exists a monomial $[x_{j_1}, x_{j_2}, \ldots, x_{j_r}, y]$, $j_i \in \{1, 2, \ldots, m\}$ with $r \leqq q$ such that*

$$[x_{i_1}, x_{i_2}, \ldots, x_{i_s}, y] \equiv [x_{j_1}, x_{j_2}, \ldots, x_{j_r}, y] \tag{1}$$

is an identical relation holding in $\mathfrak{V}^{(n)}$.

Proof. By 2.1 a $\mathfrak{V}$-free Lie algebra L freely generated by $x_1, x_2, \ldots, x_m, y$ is finite and thus the number of the monomials of the form $[x_{j_1}, x_{j_2}, \ldots, x_{j_r}, y]$ in L is finite. Taking for q the maximum of the degrees of these monomials one completes the proof of the lemma.

Lemma 2. *Assume $n \geqq m + 1$. Then there exists an $r = r(\mathfrak{V})$ such that if*

(i) *$G \in \mathfrak{V}^{(n)}$ and M is a minimal ideal of G,*

(ii) *H is a subalgebra of G such that $|G/H| \leqq m$,*

(iii) *N is an ideal of H such that $\{0\} \neq N \subseteq M$,*

then $|M| \leqq |N|^r$.

Proof. Let $x_1 + H, x_2 + H, \ldots, x_m + H$ span the k-module G/H and q be a number found in the previous lemma. Consider a k-submodule $S = \sum_{r=0}^{q} S_r$, where

$$S_l = \sum_{i_1, i_2, \ldots, i_l = 1}^{m} [x_{i_1}, x_{i_2}, \ldots, x_{i_l}, N],$$

and prove that S is a G-module. By the identities (1), $[x_i, S] \subseteq S$. To prove that S is H-closed, put $T_l = \sum_{r=0}^{l} S_r$ and notice that $T_0 = N$ and $T_q = S$. Using induction on l, we show that $[a, T_l] \subseteq T_l$ for any $a \in H$, the case $l = 0$ being trival by the condition (iii). Now for $a \in H$ one has

$$[a, T_l] = [a, T_{l-1}] + [a, S_l] \subseteq T_{l-1} + \sum_{i_1, \ldots, i_l = 1}^{m} [a, x_{i_1}, \ldots, x_{i_l}, N] \subseteq$$

$$\subseteq T_{l-1} + \sum_{i_1, i_2, \ldots, i_l = 1}^{m} [x_{i_1}, a, x_{i_2}, \ldots, x_{i_l}, N] + \sum_{i_1, i_2, \ldots, i_l = 1}^{m} [[a, x_{i_1}], x_{i_2}, \ldots, x_{i_l}, N] \subseteq$$

$$\subseteq T_{l-1} + \sum_{i_1=1}^{m} [x_{i_1}, T_{l-1}] + T_l \subseteq T_l,$$

where the Jacobi identity and the induction assumptions were used. Since $|[x_{i_1}, x_{i_2}, \ldots, x_{i_l}, N]| \leqq |N|$, $|S| \leqq |N| \sum_{l=0}^{q} m^l = |N^r|$, where r depends only on $\mathfrak{B}$. The proof is now completed by the remark that M is minimal and thus $M = S$.

Lemma 3. *Assume that $n \geqq m + 1$. Then there exists an $s = s(\mathfrak{B})$ such that if*

(i) *G is a finite algebra from $\mathfrak{B}^{(n)}$,*

(ii) *M is a minimal ideal of G such that $M, G/M \in \mathfrak{C}(f, m, c)$,*

(iii) *$|M| > m$,*

then there exists a subalgebra $H \subseteq G$, generated by at most s elements and possessing a minimal ideal N with $|N| > m$.

Proof. Let H be a minimal subalgebra of G possessing a minimal ideal N of order $> m$. Since H is the extension of the ideal $H \cap M \in \mathfrak{C}$ by the subalgebra $H + M/M \in \mathfrak{C}$ and as the orders of chief factors in $\mathfrak{C}$ are bounded by m, there exists a subalgebra (even an ideal) J in H such that $|H/J| \leqq m$. A minimal J-ideal has order $\leqq m$ and thus the previous lemma shows that $|N| \leqq m^r$. By the minimality of N for each proper k-submodule V of N there exists an $x_V \in H$ such that $[x_V, V] \nsubseteq V$. Since the number of submodules in N is evidently $< 2^{m^r}$, it suffices to put $s = 2^{m^r}$, for by the minimality of H, it is generated by all x_V and any nonzero element from N. This proves the lemma.

Now we can prove proposition 3.

Proof. If $n \geqq \max\{2, c + 1\}$ then, since all n-generator algebras from $\mathfrak{B}^{(n)}$ are in $\mathfrak{B}$ proposition 1.1 implies the validity of $(f(\operatorname{ad}(x)))\,(y) \equiv 0$ in $\mathfrak{B}^{(n)}$ and the boundedness of the nilpotency classes of the nilpotent factors by c. It suffices therefore to bound the orders of the chief factors in the finite algebras $G \in \mathfrak{B}^{(t)} \cap \mathfrak{C}^2$ by m, $t \geqq \max\{n, s\}$ (s is such as in lemma 3). Assume that G is an algebra of minimal order such that $G \notin \mathfrak{C}$. Then, by the minimality, G possesses a minimal ideal $M \neq \{0\}$ such that $|M| > m$. By our condition also $M \neq G$ and therefore G/M, $M \in \mathfrak{C}$. But in this case lemma 3 provides the existence of some s-generator subalgebra $H \subseteq G$ possessing a minimal ideal N

with $|N| > m$. This is, however, impossible by proposition 1.1, for $H \in \mathfrak{B} \subseteq \mathfrak{C}(f, m, c)$. Now the proof of the proposition is complete.

5.6. Chief centralizers and maximal subvarieties

We continue our study of the variety $\mathfrak{B} = \text{var}(A)$ generated by a finite algebra A over some finite ring k. By now we know that $\mathfrak{B}$ possesses finitely many subvarieties. Choose a maximal one: $\mathfrak{B}_0$.

Lemma 1. *Let B be a finite algebra from $\mathfrak{B}$ and $V_0(B)$ be a verbal subalgebra of B relatively to $\mathfrak{B}_0$. Then either $V_0(B) = \{0\}$ or $V_0(B)$ is a (direct) sum of minimal ideals of B.*

Proof. Clearly one may choose critical algebras $G_0, G_1, G_2, \ldots, G_s \in \mathfrak{B}$ such that $G_1, G_2, \ldots, G_s$ generate $\mathfrak{B}_0$, (QS − 1) $G_0 \in \mathfrak{B}_0$, $G_0 \notin \mathfrak{B}$ and hence $\mathfrak{B}$ is generated by a factor closed set $\mathscr{S} = \text{QS}\{G_0, G_1, \ldots, G_s\}$. If M_0 is a monolith of G_0, by the condition $V_0(G_0) = M_0$. Consider a minimal representation

$$B = C/D\,, \qquad C \subseteq P = \prod_{i=1}^{n} S_i\,, \qquad S_i \in \mathscr{S}\,, \qquad i = 1, 2, \ldots, n\,,$$

for B in $\text{var}(\mathscr{S}) = \mathfrak{B}$. Then by proposition 3.1(i) all S_i are critical. By our assumption all of them but some isomorphic copies $S_1, S_2, \ldots, S_t$ of G_0 are in $\mathfrak{B}_0$. Therefore $V_0(P) \subseteq$ $\subseteq V_0(S_1 \times S_2 \times \ldots \times S_t) = M_1 \times M_2 \times \ldots \times M_t$, $M_i \cong M_0$, are the minimal ideals of S_i, hence by proposition 3.1(iii and iv), minimal ideals of C. Thus we have $V_0(C) \subseteq$ $\subseteq T = M_1 \times M_2 \times \ldots \times M_t \subseteq C$, that is, $V_0(C)$ is an ideal of the socle of C. By proposition 1,15.1 $V_0(C)$ is a direct sum of minimal C-ideals. Consider now $V_0(C/D)$. This is an image of $V_0(C)$ under the C-homomorphism $C \to C/D$ and equal to $V_0(C) +$ $+ D/D$. By the C-homomorphism $V_0(C) + D/D = V_0(C)/V_0(C) \cap D$ and thus as a quotient module of a fully reducible C-module, $V_0(C) + D/D$ is itself fully reducible as a C- and consequently as a C/D-module. In other words $V_0(C/D)$ is a direct sum of minimal ideals of C/D which proves the lemma.

All the remaining arguments of this section will be connected with some formulae which enable to bound the indices of certain chief centralizers in algebras where they hold. They have rather bulky form and depend on two positive integers d, s and a word $v \in L(X)$. Denote them by $\mathscr{F}^v_{d,s}$. Then

$$\begin{aligned}\mathscr{F}^v_{d,s} = (\forall a, z_1, \ldots, z_l)\big[(\forall x_1, x_2, \ldots, x_d)\,\big([a, v(z_1, \ldots, z_l)] = \\ = [a, x_1, v(z_1, \ldots, z_l)] = \ldots = [a, x_1, \ldots, x_d, v(z_1, \ldots, z_l)] = 0\big) \Rightarrow \\ \Rightarrow (\forall y_1, \ldots, y_s)\,\big([a, y_1, \ldots, y_s, v(z_1, \ldots, z_l)] = 0\big)\big]\,.\end{aligned}$$

Clearly $\mathscr{F}^v_{d,s}$ with $s < d$ is identically true. The significance of such formulae for the study of centralizers is seen from the following

Lemma 2. *Assume that $\mathscr{F}^v_{s,s+1}$ holds in some Lie algebra G and let $v = v(z^0_1, \ldots, z^0_l) \in G$. Then $a \in C_G(I)$, $I = \text{id}_G(\{v\})$, the ideal generated by v in G if and only if*

$$[a, v] = [a, x_1, v] = \ldots = [a, x_1, \ldots, x_s, v] = 0\,, \tag{1}$$

for all $x_1, \ldots, x_s \in G$.

Proof. Remark first that $\mathcal{F}^v_{s,s+1}$ follows from $\mathcal{F}^v_{s-1,s}$. For if $\mathcal{F}^v_{s-1,s}$ is true in G, then to deduce $\mathcal{F}^v_{s,s+1}$ it suffices to show that, given equalities (1) in G, for any $y_1, y_2, \dots, y_{s+1} \in G$ one has $[a, y_1, \dots, y_{s+1}, v] = 0$. But by the Jacobi identity,

$$[a, y_1, y_2, \dots, y_{s+1}, v] = [y_1, a, y_2, \dots, y_{s+1}, v] + [[a, y_1], y_2, \dots, y_{s+1}, v] = \\ = [b, y_2, \dots, y_{s+1}, v], \quad (2)$$

where $b = [a, y_1]$. As $\mathcal{F}^v_{s-1,s}$ holds in G, to prove that the last member of (2) is zero, it suffices to show that $[b, x_1, x_2, \dots, x_r, v] = 0$ for all $x_1, x_2, \dots, x_r \in G$, $r < s$. But again using the Jacobi identity, $[b, x_1, x_2, \dots, x_r, v] = [[a, y_1], x_1, \dots, x_r, v] = [a, y_1, x_1, \dots, x_r, v] - [y_1, a, x_1, \dots, x_r, v] = 0$ by (1), since $r + 1 \leqq s$.

Go now on the proof of the lemma itself. The necessity of the condition (1) being evident, the sufficiency may be proved by subsequent application of the formulae $\mathcal{F}^v_{s,s+1}$, $\mathcal{F}^v_{s+1,s+2}, \dots, \mathcal{F}^v_{r-1,r}$ for any $r > s$ since any element from I is representable as a linear combination of monomials having the form $[y_1, y_2, \dots, y_r, v]$ for some $y_1, y_2, \dots, y_r$ from G. Thus the proof is complete.

Now return to $\mathfrak{V} = \operatorname{var}(A)$ and assume that $A \in \mathfrak{C}(f, m, c)$. Let also $\mathfrak{V}_0$ be a maximal proper subvariety of $\mathfrak{V}$ and $v(z_1, z_2, \dots, z_l) \equiv 0$ be an identical relation holding in $\mathfrak{V}_0$. Then one has

Lemma 3. *$\mathcal{F}^v_{m-1,s}$ is true in any algebra from $\mathfrak{V}$.*

Proof. Clearly only finitely generated and hence, by the local finiteness of $\mathfrak{V}$, only finite algebras from $\mathfrak{V}$ may be considered. Thus assume $a, z_1, \dots, z_l$ be elements of some finite algebra $H \in \mathfrak{V}$ and for all $x_1, x_2, \dots, x_{m-1}$ one has $[a, v(z_1, \dots, z_l)] = [a, x_1, v(z_1, \dots, z_l)] = \dots = [a, x_1, \dots, x_{m-1}, v(z_1, \dots, z_l)] = 0$. If $v(z_1, \dots, z_l) = 0$ then there is nothing to do. Assume then $v(z_1, \dots, z_l) \neq 0$ and using lemma 1 represent v as a sum $v = \sum_{i=1}^{t} v_i$, v_i being elements from some minimal ideals N_i of H. To prove the validity of $\mathcal{F}^v_{m-1,s}$ in H it suffices now to show that $[a, y_1, \dots, y_s, v_i] = 0$ for all $y_1, \dots, y_s \in H$. But by the proposition 1.1, $|N_i| \leqq m$, $i = 1, 2, \dots, r$, and thus the series

$$\{0\} \subseteq kv_i \subseteq [H, kv_i] + kv_i \subseteq [H, H, kv_i] + [H, kv_i] + kv_i \subseteq \dots$$

breaks at latest by its m-th term. This means, however, that any commutator of the form $[a, y_1, \dots, y_s, v_i]$ may be represented in the form of a k-linear combination of the commutators of the form $[a, x_1, x_2, \dots, x_r, v_i]$ with $r \leqq m - 1$. These latter commutators are zero by our assumptions on v and v_i, $i = 1, 2, \dots, t$, and the lemma is proved.

Now we are in a position to prove the main theorem of this chapter.

Theorem 3. *The identities of a finite Lie algebra over some finite commutative ring are finitely based.*

Proof. The theorem is proved using induction by the number of proper subvarieties of the variety $\mathfrak{V}$ generated by a finite algebra A. This is correct since on one hand the number of such subvarieties is finite and on the other hand each such subvariety is a variety generated by the direct product of its critical algebras that is by a finite algebra.

An important role is played by the following lemma 4 in which $v_0(z_1, z_2, \dots, z_l) \equiv 0$ is a defining identical relation of the variety $\mathfrak{V}_0$ which is supposed to be finitely based by the induction hypothesis. An integer g is the order of A, $g = |A|$.

Lemma 4. *Let $G \in \mathfrak{B}^{(n)}$ where $n \geqq 1 + g + \frac{(m-1)\,g(g+1)}{2}$, and H/I be a chief factor of G. If C is a centralizer of H/I in G and $H/I \notin \mathfrak{B}_0$ then $|G/C| \leqq g$.*

Proof. Of course only the case $I = \{0\}$ may be considered. Suppose thus $I = \{0\}$ and $|G/C| > g$. A desired contradiction will be obtained if we construct an identity in $\leqq n$ variables holding in $\mathfrak{B}$ but not in G. Since $H \notin \mathfrak{B}_0$, determined by $v_0 \equiv 0$, is a minimal ideal in G, for some $z_1^0, z_2^0, \ldots, z_l^0 \in H$, this ideal is generated by $v_0(z_1^0, \ldots, z_l^0) \neq 0$. Let $a_1^0, a_2^0, \ldots, a_g^0$ be elements from different classes of G by C. Now formula $\mathcal{F}_{m-1,m}^{v_0}$ is true in $\mathfrak{B}$ and includes $\leqq n$ variables. Therefore it holds in $\mathfrak{B}^{(n)}$ and thus by lemma 2 there exist $x_1^0, \ldots, x_r^0 \in G$, $0 \leqq r \leqq m-1$, such that $0 \neq [a_1^0, x_1^0, \ldots, x_r^0, v_0(z_1^0, \ldots, z_l^0)]$ and therefore the identical relation $v_1(a_1; x_1, \ldots, x_r; z_1, \ldots, z_l) = [a_1, x_1, \ldots, x_r, v(z_1, \ldots, z_l)] \equiv 0$, holding in $\mathfrak{B}_0$ fails in G. Using next $\mathcal{F}_{m-1,m}^{v_1}$ one can find $x_{r+1}^0, \ldots, x_{r+s}^0$ in G, $0 \leqq s \leqq m-1$, such that

$$[a_2^0 - a_1^0, x_{r+1}^0, \ldots, x_{r+s}^0, v_1(a_1^0; x_1^0, \ldots, x_r^0; z_1^0, \ldots, z_l^0)] \neq 0\,.$$

Proceeding similarly one obtains finally a word

$$\bar{v} = v_{g(g+1)/2}\,(a_1, \ldots, a_g; x_1, \ldots, x_d; z_1, \ldots, z_l)$$

depending on $\leqq n$ variables a_i, x_j, z_s, which is not identical zero in G and includes as its factors all variables $a_1, \ldots, a_g$ and their differences $a_i - a_j$, $i > j$. At every step of its construction one applies the formulae $\mathcal{F}_{m-1,m}^{v_s}$, $0 \leqq s < \frac{g(g+1)}{2}$, which hold in $\mathfrak{B}$ and, depending on $\leqq n$ variables, are true also in $A \in \mathfrak{B}^{(n)}$.

On the other hand $\bar{v} \equiv 0$ is an identity in A, for $|A| = g$ implies that either values of some a_i equal that of some a_j for $i > j$, or the value of some a_i equals zero. Thus we finished the proof of the lemma.

Return now to the proof of the main theorem. Let s be an integer found in the lemma 5.3, n be that of the preceding lemma and $t = \max\{s, n\}$. Using the corollary to proposition 5.2 it suffices to show that $\mathfrak{B}^{(t)} \subseteq \mathfrak{C}(f, m, c)$, if $\mathfrak{B} \subseteq \mathfrak{C}(f, m, c)$. Since by theorem 4.2 $\mathfrak{C}$ is a variety, we may prove only that any finitely generated algebra $G \in \mathfrak{B}^{(t)}$ belongs to $\mathfrak{C}$. Consider $C(G)$, the verbal subalgebra of G corresponding to $\mathfrak{C}$, and assume $C(G) \neq \{0\}$. By our assumption and proposition 1.2, $G/C(G)$ is finite and therefore $H = C(G)$ is finitely generated as an ideal of G. Consequently there exists a maximal ideal I of G contained in H with $I \neq H$. To obtain a contradiction it suffices to show that $G/I \in \mathfrak{C}$. Therefore we assume without loss of generality $I = \{0\}$, that is H is a minimal ideal of G. We prove that H is a finite algebra from $\mathfrak{C}$. For, by theorem 4,9.9, H is finitely generated as a subalgebra and if $H \in \mathfrak{B}_0 \subset \mathfrak{C}$ then H is finite. If C is the centralizer of H in G then either $C \supseteq H$ and then $H \in \mathfrak{C}$ (for $H \in \mathfrak{B}^{(n)} \subseteq \subseteq \mathfrak{B}^{(1)}$ in this case implies $H \in \mathfrak{B} \subseteq \mathfrak{C}$), or $C \cap H = \{0\}$ and $|H| \leqq |G/C|$. If $H \notin \mathfrak{B}_0$, by the previous lemma $|G/C| \leqq g$, which implies $H \in \mathfrak{B}$ (for $H \in \mathfrak{B}^{(n)} \subseteq \mathfrak{B}^{(g)}$).

Finally, G is a finite algebra from $\mathfrak{B}^{(n)} \cap \mathfrak{C}^2$, $n \geqq S$, so that the proposition 5.3 applies and $G \in \mathfrak{C}$. This contradiction (on the other hand $C(G) = H \neq \{0\}$) completes the proof of the main theorem.

Corollary. *$\mathfrak{C}(f, m, c)$ is a finitely based variety.*

CHAPTER 6

Applications to the group theory

6.1. Topological algebras

An algebra T is called a *topological* one if T is a topological space and all operations of T are continuous in the corresponding topology. For example, if T is endowed by bracketing and, for some $x_0, y_0, z_0 \in T$, $[x_0, y_0] = z_0$, then for any open set $U_3 \ni z_0$ there exist open sets $U_1 \ni x_0$, $U_2 \ni y_0$ such that for any $x \in U_1$, $y \in U_2$ one has $[x, y] \in U_3$. All algebras to be considered will possess a countable basis for neighbourhoods of zero and therefore the completeness of the space T may be defined through Cauchy sequences. Namely, a sequence $\{a_n\}_{n\geq 1}$ of elements of T is called a *Cauchy sequence* if, for any open set $V \ni 0$, there exists a positive integer N such that, for any $s, t > N$, $a_s - a_t \in V$. A sequence $\{a_n\}_{n\geq 1}$ *converges* to the element $a \in T$ if, for any open set $V \ni 0$, there exists a positive integer N such that, for any $s > N$, $a_s - a \in V$. We denote this by $\{a_n\} \to a$. A *complete* space is such one in which any Cauchy sequence converges. To explain the role of complete topological algebras, recall that the map f: $T \to T'$ of two topological spaces is called continuous if, given $x_0 \in T$, $y_0 \in T'$ with $y_0 = f(x_0)$, for any open set $V \ni y_0$, there exists an open set $U \ni x_0$ such that for any $x \in U, f(x) \in V$. Say also that a subset S is *dense* in T if its closure coincides with T. Now if T' is a complete separable topological algebra then any continuous homomorphism f of S into T' extends to the unique continuous homomorphism $\bar{f}$ of T into T'. For, by our condition any element $a \in T$ is representable in the form $\{a_n\} \to a$, $a_n \in S$, $n = 1, 2, \ldots$. Now it suffices to notice that $\{f(a_n)\}_{n\geq 1}$ is a Cauchy sequence in T' and hence converges to some definite element $\bar{f}(a)$ which now defines $\bar{f}$ for any $a \in T$, the continuity of $\bar{f}$ following from the well known regularity of separable topological groups.

A topology $\mathcal{T}_1$ is *stronger* then $\mathcal{T}_2$ (of the same space T) if every open set $V \in \mathcal{T}_2$ contains some open set $U \in \mathcal{T}_1$. Every Cauchy sequence with respect to $\mathcal{T}_2$ is so with respect to $\mathcal{T}_1$. An identity map of T endowed by $\mathcal{T}_1$ into T endowed by $\mathcal{T}_2$ is continuous. We shall use this remark when considering completions of a topological algebra. Namely, call the complete separable algebra $\hat{T}$ a *separable completion* of T if there exists a continuous homomorphism φ: $T \to \hat{T}$ such that, given a continuous homomorphism ψ: $T \to T'$, T' being some other separable complete topological algebra, there exists a unique continuous homomorphism χ: $\hat{T} \to T'$ such that $\chi\varphi = \psi$. Such a completion always exists and is unique up to continuous isomorphism, the proof in the particular case interesting for us to be given below. Note here only that by the remark above if $\hat{T}$, $\tilde{T}$ are two separable completions with respect to separable topologies $\mathcal{T}_1, \mathcal{T}_2, \mathcal{T}_1$ being stronger than $\mathcal{T}_2$, then the canonical map φ_2 of $\hat{T}$ into $\tilde{T}$ induces a canonical injection of

$\hat{T}$ into $\tilde{T}$. Our particular case is that of so called *filtered* topological algebras. Namely, let T be an algebra and $\mathcal{F}$ be the (countable) set of its ideals. Consider the weakest topology in which all sets of the form $t + F$, $F \in \mathcal{F}$ are open. Then all $F \in \mathcal{F}$ being ideals implies $[t_1 + F_1, t_2 + F_2] \subseteq [t_1, t_2] + F_1 \cup F_2$ and using this one immediately observes that relatively to this topology T becomes a topological algebra. This algebra is separable if and only if the intersection of the set $\mathcal{F}$ is zero (for then $t_1 - t_2 \notin F$ implies $t_1 + F \cap t_2 + F = \emptyset$ that is the existence of two disjoint neighbourhoods containing t_1, t_2 respectively).

The completion of such an algebra may be constructed using the concept of a *projective limit* of algebras (or groups, etc.). Namely, given a family $(T^{(\alpha)})_{\alpha \in I}$ of, say, algebras, I being partially ordered, and their homomorphisms $\varphi_{\alpha\beta}\colon T^{(\alpha)} \to T^{(\beta)}$, $\alpha > \beta$, consider a cartesian product $P = C\prod_{\alpha \in I} T^{(\alpha)}$ and its subset $\tilde{T}$ formed by functions f with $\alpha > \beta$ implies $\varphi_{\alpha\beta}(f(\alpha)) = f(\beta)$. Clearly, $\tilde{T}$ is a subalgebra, called the *projective limit* of the family $(T^{(\alpha)}, \varphi_{\alpha\beta})$ and denoted by $\tilde{T} = \varprojlim T^{(\alpha)}$. Given a set of elements $(x^\alpha)_{\alpha \in I}$, $x^\alpha \in T^\alpha$, with $\varphi_{\alpha\beta}(x^\alpha) = x^\beta$ (or simply $\varphi(x^\alpha) = x^\beta$), denote by $\varprojlim x^\alpha$ a function $f \in \tilde{T}$ such that $f(\alpha) = x^\alpha$. In the general case $\tilde{T}$ may well happen to consist only of zero (or unity, in the group case, etc.) function.

We consider, however, the case of $\mathcal{F}$ being a decreasing chain of ideals T_n, $n \geqq 1$, and denote by $T^{(n)}$ the quotient algebras $T^{(n)} = T/T_n$ and by $\varphi_{nm}\colon T^{(n)} \to T^{(m)}$, $n > m$, the epimorphisms induced by the natural epimorphisms $\varepsilon_n\colon T \to T/T_n = T^{(n)}$, $n, m = 1, 2, \ldots$ Let $\tilde{T}$ be the projective limit of the family $\{T^{(n)}, \varphi_{nm}\}_{n,m \geq 1}$ and $\tilde{T}_m$ be an *ideal* of elements $x = \varprojlim x^n$ with $x_1 = \ldots = x_m = 0$. Clearly $\tilde{T}_1 \supseteq \tilde{T}_2 \supseteq \ldots$, $\bigcap_{m=1}^{\infty} \tilde{T}_m = \{0\}$ and therefore the topology of $\tilde{T}$ determined by $\{\tilde{T}_m\}_{m \geq 1}$ is separable. We prove that $\tilde{T}$ is complete. Let $\{x_n\}_{n \geq 1}$ be a Cauchy sequence of elements $x_n = \varprojlim x_n^m$. By the definition, for any $m \geqq 1$, there exists an integer $c(m)$ $(> c(m-1))$ such that for any $n_1, n_2 \geqq c(m)$, $x_{n_1} - x_{n_2} \in \tilde{T}_m$. Considering then the element $x = \varprojlim x^n_{c(n)}$ we prove that $x \in \tilde{T}$ and $\{x_n\} \to x$. Indeed, it suffices to show first that $\varphi(x^{n+1}) = x^n$. But $x^n = x^n_{c(n)} = x^n_{c(n+1)}$ and hence $\varphi(x^{n+1}_{c(n+1)}) = x^n_{c(n+1)} = x^n_{c(n)}$. Now the reader will certainly be able to prove $\{x_n\} \to x$ by himself. Remark next, that under any continuous homomorphism ψ of T into a separable algebra, $\psi(T_\infty) = \{0\}$, where $T_\infty = \bigcap_{n=1}^{\infty} T_n$. Therefore to prove that $\tilde{T}$ is a separable completion of T it remains to note that the mapping $i\colon T \to \tilde{T}$ given by $i(x) = \varprojlim \varepsilon_n(x)$ induces an isomorphism of T/T_∞ such that the topologies of $i(T)$ induced by those of T and $\tilde{T}$ coincide, and $i(T)$ is dense in T. The first of these statements is however clear from $i(T) \cap \tilde{T}_n = i(T_n)$ and the second is proved by taking for any $x \in \tilde{T}$ a set $y_n \in T$ with $\varepsilon_n(y_n) = x_n$ and showing that $\{y_n\} \to x$.

From the above argument one immediately deduces such a remark: if T belongs to some variety of algebras, so does its completion $\tilde{T}$. Thus, for example, if T is an associative or Lie algebra, then the same is true for $\tilde{T}$.

Examples. (i) Let $A = A(X)$ be the free associative algebra with unity over k having X as the set of free generators; denote by $A_{(n)}$ the sum $\bigoplus_{s \geq n} A_s(X)$ of all homogeneous components of A of degree $\geqq n$, $n = 0, 1, 2, \ldots$. Then the filter $\{A_{(n)}\}$, $n \geqq 0$, determines a topology of A, while the completion $\tilde{A}$ in this particular case may be found in the following form. Consider $\hat{A} = C \prod_{n \geqq 0} A_n$ and define the multiplication in $\hat{A}$ by the rule $(f, g \in \hat{A})$: $(fg)(n) = \sum_{i=0}^{n} f(i)\, g(n - i)$. If $\hat{A}_{(n)} = C \prod_{s \geqq n} A_s$ then $\{\hat{A}_{(n)}\}$ is a basis of neighbourhoods of zero, $\hat{A}$ being a complete topological algebra with respect to this topology. An embedding i of A into $\hat{A}$ is given by $i(a) = f_a$, where $f_a(n) = a_n$, the n-th homogeneous component of a, $n = 1, 2, \ldots$. Since $i(A_{(n)}) = i(A) \cap \hat{A}_{(n)}$ and $i(A)$ is dense in $\hat{A}$, this latter is a completion of A.

(ii) Let, in particular, A be a polynomial ring in one variable t, $A = k[t]$. Then $\hat{A} = k[[t]]$ is a formal power series ring in t over k.

(iii) An algebra $\hat{A}(X)$, called sometimes a *Magnus algebra of a set X*, contains a Lie subalgebra $\hat{L}(X)$ formed by the functions f such that $f(n) \in L_n \subseteq A_n$, $n = 0, 1, 2, \ldots$. Again $\hat{L}(X)$ is a completion of $L(X)$.

(iv) An argument of the example (i) is valid for any graded algebra without even minor changes. Let further B, C be two graded algebras and $D = B \otimes C$ be their tensor product, that is the tensor product of the underlaying vector spaces with multiplication given by $(b_1 \otimes c_1)(b_2 \otimes c_2) = b_1 b_2 \otimes c_1 c_2$. Denote by D_n the subspace

$$B_0 \otimes C_n \oplus B_1 \otimes C_{n-1} \oplus \ldots \oplus B_{n-1} \otimes C_1 \oplus B_n \otimes C_0$$

of the algebra D. The reader immediately sees that $(D_n)_{n \geqq 0}$ is a graduation of D. Denote then by $\hat{D}$ (or by $\widehat{B \otimes C}$) the completion of D constructed in the example (i). Clearly $\widehat{B \otimes C}$ contains a subalgebra isomorphic to $\hat{B} \otimes \hat{C}$ but does not in general coincide with this latter.

If T is a complete separable topological algebra then one may consider *infinite sums* of its elements in the following sense: put $x = \sum_{n=1}^{\infty} x_n$ if, denoting by the s_m the sum $s_m = x_1 + x_2 + \ldots + x_m$, $m = 1, 2, \ldots$, one has $\{s_m\} \to x$. In the filtered algebras a necessary and sufficient condition for a sum $\sum x_n$ to converge is $\{x_n\} \to 0$. For then, given $F \in \mathcal{F}$, the generating filter of open sets, there exists an N such that $x_n \in F$ for all $n \geqq N$. But obviously in this case any sum $x_n + x_{n+1} + \ldots + x_m = s_m - s_n \in F$, $m > n > N$ and $\{s_m\}$ is a Cauchy sequence. By the completeness of T for some $x \in T$, $\{s_m\} \to x$. In particular, any element $x \in A$ has the form $x = \sum_{n=0}^{\infty} x_n$, $x_n \in A_n$, and $x \in \hat{L}$ if and only if $x_n \in L_n$, $n = 0, 1, \ldots$.

We conclude this section by a criterion which enables to determine, given an element $a \in \hat{A}(X)$, whether $a \in \hat{L}(X)$ or not. To do this we need first a

Lemma. *Let $\Delta: A(X) \to A(X) \otimes A(X)$ be a homomorphism extending the mapping $x \mapsto x \otimes 1 + 1 \otimes x$, $x \in X$. Then $a \in A = A(X)$ is an element of $L = L(X)$ if and only if $\Delta(a) = a \otimes 1 + 1 \otimes a$.*

Proof. Using the above introduced graduation of $A \otimes A$ we immediately osberve that one may without loss of generality limit oneself to the case of homgoeneous elements. The sufficiency then easily follows, since if $a \mapsto a \otimes 1 + 1 \otimes a$ and $b \mapsto$

$\mapsto b \otimes 1 + 1 \otimes b$, then $[a, b] \mapsto [a \otimes 1 + 1 \otimes a,\ b \otimes 1 + 1 \otimes b] = [a, b] \otimes 1 + 1 \otimes [a, b]$, $1 \otimes u$ and $v \otimes 1$ commuting for all $u, v \in A$.

Suppose then, conversely, $\Delta(a) = a \otimes 1 + 1 \otimes a$, $a \in A$. Choose a totally ordered basis W in L, and recall that by the PBW-theorem a 2,4.6 a basis of A is formed then by monomials $b = w_1 w_2 \dots w_s$, where $w_1 \geqq w_2 \geqq \dots \geqq w_s$. Then $\Delta(b) = (w_1 \otimes 1 + 1 \otimes w_1) \times (w_2 \otimes 1 + 1 \otimes w_2) \dots (w_s \otimes 1 + 1 \otimes w_s)$ by the first part of this proof. Clearly, $\Delta(b) = \sum_{r+t=s} w_{i_1} \dots w_{i_t} \otimes w_{j_1} \dots w_{j_r}$, and $i_1 \geqq \dots \geqq i_t$, $j_1 \geqq \dots \geqq j_r$, that is $\Delta(b)$ is written as a sum of basic elements of $A \otimes A$, if the canonical basis of the tensor product is used. Let then $b' = w_1' w_2' \dots w_{s'}'$, $w_1' \geqq w_2' \geqq \dots \geqq w_{s'}'$. Assume that $\Delta(b)$ and $\Delta(b')$ have equal members in their expression through the basic elements of $A \otimes A$. Then one has $w_{i_1} \dots w_{i_t} \otimes w_{j_1} \dots w_{j_r} = w_{m_1}' \dots w_{m_{t'}}' \otimes w_{l_1}' \dots w_{l_{r'}}'$. Clearly, $t = t'$, $r = r'$ and therefore $s = s'$. Moreover the sets $\{w_1, \dots, w_s\}$ and $\{w_1', \dots, w_s'\}$ then coincide; since, further, they are ordered, $b = w_1 \dots w_s = w_1' \dots w_s' = b'$.

This yields now that if a includes in its basic expression monomials b with $s > 1$ then $\Delta(a)$ includes in its basic expression monomials different from $w_1 w_2 \dots w_s \otimes 1$ and $1 \otimes w_1 w_2 \dots w_s$ and therefore $\Delta(a) \neq a \otimes 1 + 1 \otimes a$. Hence $s = 1$ and thus $a \in L$. This completes our proof.

Note now that relatively to topologies in A and $A \otimes A$ introduced in the examples (i) and (iv), the homomorphism Δ is continuous (this is true for any *homogeneous*, that is graduation preserving mapping). Therefore there exists a continuous homomorphism $\hat{\Delta}: \hat{A} \to \widehat{A \otimes A}$ such that $\Delta\left(\sum_{n=0}^{\infty} a_n\right) = \sum_{n=0}^{\infty} \Delta(a_n)$. Thus if $a = \sum_{n=0}^{\infty} a_n \in \hat{L}$, that is, each $a_n \in L_n$, then $\hat{\Delta}(a) = \sum_{n=0}^{\infty} \Delta(a_n) = \sum_{n=0}^{\infty} (a_n \otimes 1 + 1 \otimes a_n) = \left(\sum_{n=0}^{\infty} a_n\right) \otimes 1 + 1 \otimes \left(\sum_{n=0}^{\infty} a_n\right) = a \otimes 1 + 1 \otimes a$. If conversely $\hat{\Delta}(a) = a \otimes 1 + 1 \otimes a$ then by the same argument $a_n \in \hat{L}_n$ for all $n = 0, 1, 2, \dots$, hence $a \in \hat{L}$. So we have proved

Proposition 1. *An element $a \in \hat{A}$ is an element of $\hat{L}$ if and only if $\hat{\Delta}(a) = a \otimes 1 + 1 \otimes a$, $\hat{\Delta}$ being the continuous homomorphism of $\hat{A}$ into $\widehat{A \otimes A}$ extending the mapping $x \mapsto x \otimes 1 + 1 \otimes x$, $x \in X$.*

6.2. Campbell-Hausdorff formula

Let $k[[t]]$ be a formal power series ring, k being a zero characteristic field, and e^t, $\ln(1 + t)$ be two elements given by

$$e^t = \sum_{n=0}^{\infty} \frac{t^n}{n!}, \qquad \ln(1 + t) = \sum_{n=1}^{\infty} (-1)^{n-1} \frac{t^n}{n}.$$

Let further B be a complete topological algebra and $\varphi: k[[t]] \to B$ be a continuous homomorphism. Then $\varphi(e^t) = \varphi\left(\sum_{n=0}^{\infty} \frac{t^n}{n!}\right) = \sum_{n=0}^{\infty} \frac{(\varphi(t))^n}{n!}$ is a well determined element of B denoted by $e^{\varphi(t)}$. Similarly one defines $\ln \varphi(t)$. By direct calculations one may verify that $t \mapsto \ln(i + t)$ extends to a continuous homomorphism of $k[[t]]$ into itself and that $e^{\ln(1+t)} = 1 + t$. Similarly, $\ln e^t = t$.

The reader is suggested to verify that, given an element $a \in \hat{A}(X)$ without constant term, the mapping $t \mapsto a$ extends to a continuous homomorphism of $k[[t]]$. Denote now by M the ideal consisting of all such a. Let also $G = G(X) = 1 + M$ be the set of series with constant term 1. Then G is a group under multiplication, for $(1 + a)^{-1} = 1 - a + a^2 - \dots$ and the mappings $a \mapsto e^a$, $1 + a \mapsto \ln(1 + a)$ by the previous remarks are one-to-one mutually reciprocal mappings of M onto G and G onto M respectively. We consider also $e^{\hat{L}}$, which is correct since $\hat{L} \subseteq M$.

Theorem 1. $e^{\hat{L}}$ *is a subgroup of* G.

Proof. Since $e^a e^b = e^c$ for any $a, b \in \hat{L}$ and some unique $c \in M$, it suffices to prove that $c \in \hat{L}$. To do this note first that if a and b commute then, obviously, $e^a e^b = e^{a+b}$, thus $c = a + b \in L$ and we are finished. In the general case, write $c = \ln(e^a e^b)$ and apply to both sides of this equality the continuous *homomorphism* $\hat{\Delta}$ of the proposition 1.1. It is easily then seen (we omit some intermediate details) that

$$\begin{aligned}\hat{\Delta}(c) &= \hat{\Delta}(\ln(e^a e^b)) = \ln(e^{\hat{\Delta}(a)} e^{\hat{\Delta}(b)}) = \ln(e^{a\otimes 1 + 1\otimes a} e^{b\otimes 1 + 1\otimes b}) = \\ &= \ln e^{a\otimes 1} e^{1\otimes a} e^{b\otimes 1} e^{1\otimes b} = \ln(e^a e^b \otimes 1)(1 \otimes e^a e^b) = \\ &= \ln(e^a e^b \otimes 1) + \ln(1 \otimes e^a e^b) = (\ln(e^a e^b)) \otimes 1 + 1 \otimes (\ln(e^a e^b)) = \\ &= c \otimes 1 + 1 \otimes c\,.\end{aligned}$$

By the proposition 1.1, $c \in \hat{L}$ and the theorem follows.

Let next $X = \{x, y\}$, $\hat{A} = \hat{A}(X)$, $\hat{L} = \hat{L}(X)$, and consider the unique element $x \circ y \in \hat{L}$ such that $e^x e^y = e^{x \circ y}$. Given now some complete topological Lie algebra B over k the topology of B being determined by a decreasing chain of ideals B_n with B/B_n nilpotent, one may introduce in B a new binary operation $\circ$ by putting $b \circ c = \varphi(x \circ y)$, φ being a continuous homomorphism of $\hat{L}$ into $\hat{B}$ extending the mapping $x \mapsto b$, $y \mapsto c$.

Proposition 1. $\langle B, \circ \rangle$ *is a group.*

Proof. First remark that the homomorphism $\varphi\colon L \to B$ extending the above mentioned mapping is continuous, for since $\mathrm{alg}(\{x, y\}) + B_n/B_n$ is nilpotent of class, say, $c(n)$, $\varphi(L_{c(n)+1}) = \varphi(L^{c(n)+1}) \subseteq B_n$ for any $n = 1, 2, \dots$. Therefore, φ extends to a continuous homomorphism $\hat{\varphi}$ of $\hat{L}$ into B.

The group axioms for $\circ$ may be verified only in the case $B = \hat{L}(\{x, y, z\}) \subseteq \hat{A}(\{x, y, z\})$. Since $x \circ 0 = 0 \circ x = x$ and $x \circ (-x) = (-x) \circ (x) = 0$, it suffices to show that $x \circ (y \circ z) = (x \circ y) \circ z$. But $(x \circ y) \circ z = \ln e^{(x\circ y)\circ z} = \ln e^{x\circ y} e^z = \ln(e^x e^y) e^z = \ln e^x(e^y e^z) = \ln e^x e^{y\circ z} = \ln e^{x\circ(y\circ z)} = x \circ (y \circ z)$ since exponentials are elements of the *associative* algebra $\hat{A}$. The proof is thus complete.

We call $\langle B, \circ \rangle$ "*a group of a Lie algebra B*" and denote it by B°.

Remarks. (i) Let $C \subseteq B$ be a closed subalgebra of B. Then C is a subgroup of B, moreover this subgroup is identically isomorphic with C°. We shall denote it also by C°.

(ii) Let $\varphi\colon B \to D$ be a continuous epimorphism of complete Lie algebras B and D. Then φ preserves multiplication and therefore is a homomorphism of groups B° onto D°. We denote it by φ°.

(iii) Since each closed ideal C may be considered as a kernel of some continuous homomorphism $\varphi\colon B \to B/C$ and, by (ii), $\operatorname{Ker}\varphi = \operatorname{Ker}\varphi^\circ$ is a normal subgroup of B°. Applying once more the homomorphism theorem, one has $B^\circ/C^\circ \cong (B/C)^\circ$.

It is useful often to have an explicit form for $x \circ y$ as a series in x, y. This can be done using one more criterion for some element from $A(X)$ to belong to $L(X)$. Define the linear map $\theta\colon A^+(X) \to L(X)$ by the rule $\pi(x_1 x_2 \ldots x_n) = (\mathrm{ad}(x_1))\,(\mathrm{ad}(x_2)) \ldots (\mathrm{ad}(x_{n-1}))\,(x_n)$, $x_1, x_2, \ldots, x_n \in X$.

Then the following proposition holds:

Proposition 2. *An element $a \in A_n(X)$ is an element of $L_n(X)$ if and only if $\pi(a) = na$.*

Proof. This condition is vacuously sufficient. Prove its necessity. Consider L as a natural left A-module and denote the action by $a * w$, $a \in A$, $w \in L$. Thus $w_1 * w = (\mathrm{ad}(w_1))\,(w)$ for all $w_1 \in L$. Evidently from the definition of π, $\pi(ab) = a * \pi(b)$ for all $a \in A$, $b \in A^+$. We show that π is a derivation of L. Indeed, $\pi([u, v]) = \pi(uv - vu) = u * \pi(v) - v * \pi(u) = [u, \pi(v)] + [\pi(u), v]$. We proceed finally by induction on n; since $\pi(a) = a$ for $a \in L_1(X) = A_1(X)$, assume that $a = [u, v]$ is a monomial from $L_n(X)$, $u \in L_l$, $v \in L_m$, $l + m = n$. Then by the induction hypothesis, $\pi(a) = \pi([u, v]) = [\pi(u), v] + [u, \pi(v)] = [lu, v] + [u, mv] = na$, and the proof is complete.

The promised explicit form for $x \circ y$ is now obtained by the following calculation

$$x \circ y = \ln(e^x e^y) = \ln\left(1 + \sum_{n+m \geq 1} \frac{x^n y^m}{n!\,m!}\right) = \sum_{t>0} (-1)^{t-1} \frac{1}{t} \left(\sum_{n+m \geq 1} \frac{x^n y^m}{n!\,m!}\right)^t =$$
$$= \sum_{t>0} (-1)^{t-1} \frac{1}{t} \sum_{n_i + m_i \geq 1} \frac{x^{n_1} y^{m_1} x^{n_2} y^{m_2} \ldots x^{n_t} y^{m_t}}{n_1!\,m_1!\,n_2!\,m_2! \ldots n_t!\,m_t!}. \tag{1}$$

Let $(x \circ y)_s$ be the homogeneous component of the degrees in x, y. Then by the preceding proposition and (1), one observes that $s(x \circ y)_s$ is a linear combination of monomials of the form

$$(\mathrm{ad}(x))^{n_1}\,(\mathrm{ad}(y))^{m_1} \ldots (\mathrm{ad}(x))^{n_t}\,(\mathrm{ad}(y))^{m_t - 1}\,(y)$$

if $m_t > 0$ and of the form

$$(\mathrm{ad}(x))^{n_1}\,(\mathrm{ad}(y))^{m_1} \ldots (\mathrm{ad}(x))^{n_t - 1}\,(x)\,,$$

if $m_t = 0$. Now we have the *Campbell-Hausdorff formula*

$$x \circ y = \sum_{s=1}^{\infty} \frac{1}{s} \sum_{t=1}^{m} \frac{(-1)^{t-1}}{t} \left(\sum_{\substack{n_1 + \ldots + n_t = p \\ m_1 + \ldots + m_t = q \\ p+q=s,\, n_i + m_i \geq 1}} \frac{(\mathrm{ad}(x))^{n_1} \ldots (\mathrm{ad}(y))^{m_t - 1}\,(y)}{n_1! \ldots m_t!} + \right.$$
$$\left. + \sum_{\substack{n_1 + \ldots + n_t = p \\ m_1 + \ldots + m_{t-1} = q \\ p+q=s,\, n_i + m_i \geq 1}} \frac{(\mathrm{ad}(x))^{n_1} \ldots (\mathrm{ad}(y))^{m_{t-1}}\,(\mathrm{ad}(x))^{n_t - 1}\,(x)}{n_1! \ldots m_{t-1}!\,n_t!} \right). \tag{2}$$

In particular, $(x \circ y)_0 = 0$, $(x \circ y)_1 = x + y$, $(x \circ y)_2 = \frac{1}{2}[x, y]$, that is

$$x \circ y = x + y + \tfrac{1}{2}[x, y] + \ldots$$

6.3. Free groups

Let X and X^{-1} be two nonempty sets and $x \mapsto x^{-1}$ be a bijection of X onto X^{-1}. Consider a semigroup $W = W(X \cup X^{-1})$ of (possibly empty) words in $X \cup X^{-1}$ relatively to juxtaposition. Denote the empty word by 1 and say that two words u and v are *equivalent*, written as $u \sim v$, if u can be obtained from v by a finite (possibly zero) number of insertions or deletions of subwords having the form xx^{-1} or $x^{-1}x$ for some $x \in X$. It is easily seen that this relation is an equivalence one, and we denote its classes by square brackets $[u]$, u being some representative from this class. A word u is called *irreducible* if u does not contain any subword of the above mentioned form.

Lemma. *Any class* $[u]$ *contains only one irreducible word.*

Proof. Denote by $\varrho(u)$ a single irreducible word obtained from u by subsequent deletions of subwords of the form $x^\varepsilon x^{-\varepsilon}$ (that is xx^{-1} if $\varepsilon = 1$ and $x^{-1}x$ if $\varepsilon = -1$) from the right to the left. Then clearly $\varrho(u) \sim u$ and $\varrho(u) = u$ if u is irreducible. Next, $\varrho(uv) = \varrho(u\varrho(v))$. This yields $\varrho(ux^\varepsilon x^{-\varepsilon}v) = \varrho(uv)$, $\varepsilon = \pm 1$, and finally, proceeding by induction over the length of u, one has $\varrho(uv) = \varrho(u)\,\varrho(v)$. Let now $u \sim v$ be two irreducible words. Then by the definition, there exists a sequence

$$u = u_1, u_2, \dots, u_m = v$$

such that u_{i+1} is obtained from u_i by one insertion or one deletion of a subword having the form $x^\varepsilon x^{-\varepsilon}$, $\varepsilon = \pm 1$, $x \in X$. By our formulae then $\varrho(u_{i+1}) = \varrho(u_i)$. This, however, implies $\varrho(u) = \varrho(v)$ and, since u, v are irreducible, $u = v$. The proof of the lemma is then complete.

Now we introduce an operation on the set $F(X)$ of classes of words by putting $[u]\,[v] = [uv]$. This definition is unambiguous since if $u \sim u'$, $v \sim v'$ then $\varrho(u) = \varrho(u')$, $\varrho(v) = \varrho(v')$ and using $\varrho(uv) = \varrho(\varrho(u)\,\varrho(v))$ gives $\varrho(uv) = \varrho(u'v')$, hence $uv \sim u'v'$. The associativity of $F(X)$ follows from that of $W(X \cup X')$; moreover $[1]$ is a unity of $F(X)$ and $[u^{-1}]$ is an inverse of $[u]$, u^{-1} being defined as the image of u under the homomorphism of W induced by $x \mapsto x^{-1}$, $y^{-1} \mapsto y$, $x, y \in X$. Therefore one has

Proposition 1. $F(X)$ *is a group generated by all* $[x]$, $x \in X$. *Moreover, given another group* G *and a mapping* $\varphi\colon X \to G$, *there exists a unique homomorphism* $\chi\colon F(X) \to G$ *such that* $\chi([x^\varepsilon]) = (\varphi(x))^\varepsilon$.

Proof. Only the second statement needs verification. But consider the mapping $\chi\colon X \cup X^{-1} \to G$ defined by $\chi(x^\varepsilon) = (\varphi(x))^\varepsilon$, $\varepsilon = \pm 1$. Then there exists a unique homomorphism $\bar{\chi}$ (of semigroups) of W into G extending χ. Its kernel Ker χ evidently contains $\sim$, so that χ induces a unique homomorphism $\psi\colon F = W/\!\sim\ \to G$ which is clearly a homomorphism of groups with the required properties.

We omit brackets in the expression of elements from F, this latter group being called a *free group* with set X of free generators. Thus any element of F is uniquely representable in the form $x_1^{\varepsilon_1}x_2^{\varepsilon_2} \dots x_n^{\varepsilon_n}$, $x_i \in X$, $\varepsilon_i = \pm 1$ and $\varepsilon_i \neq -\varepsilon_{i+1}$, $i = 1, 2, \dots, n$.

Corollary. *Every group is a quotient group of a free group.*

Indeed, given any group G, consider a set X with $|X| = |G|$. Then by the proposition, the bijection $X \to G$ extends to a homomorphism and hence to the epimorphism of $F = F(X)$ onto G.

Many groups are known to be free, for example the reader may easily verify that the subgroup of a group of nondegenerate integer matrices generated by $x = \begin{pmatrix} 1 & 2 \\ 0 & 1 \end{pmatrix}$ and $y = \begin{pmatrix} 1 & 0 \\ 2 & 1 \end{pmatrix}$ is free. We show now that the group $G = G(X) = 1 + M \subseteq \hat{A}(X)$ also possesses a number of them. Namely, the following theorem holds:

Theorem 2. *Any set of elements a_x, $x \in X$, having the form*

$$a_x = 1 + x + \gamma(x)\,, \qquad \gamma(x) \in \hat{A}_{(2)}(X)\,,$$

generate a free subgroup of $G(X)$ isomorphic to $F(X)$, the base ring k being an arbitrary commutative ring with 1.

Proof. Assume first $\gamma(x) = 0$ for all $x \in X$. We define a homomorphism of F into G extending the mapping $x \mapsto a_x$, $x \in X$, and, denoting the image of $b \in F$ under this homomorphism by $\bar{b}$, show that $b \neq 1$ implies $\bar{b} \neq 1$. So far, we represent b in the form $b = x_1^{m_1} x_2^{m_2} \dots x_n^{m_n}$, $m_i \neq 0$, $x_i \neq x_{i+1}$, $i = 1, 2, \dots, n$. Then $\bar{b} = (1 + x_1)^{m_1} (1 + x_2)^{m_2} \dots \dots (1 + x_n^{m_n})$. Since by the binomial formula $(1 + x_i)^{m_i} = 1 + m_i x_i + n_i x_i^2 + \dots$, $i = 1, 2, \dots, n$, the component of the degree n in b contains $x_1 x_2 \dots x_n$ with the coefficient $m_1 m_2 \dots m_n \neq 0$. This proves $\bar{b} \neq 1$ and the freeness of the subgroup $H = \mathrm{gp}(\{1 + x, x \in X\})$.

The general case is dealt with by considering a continuous homomorphism $\varphi\colon \hat{A}(X) \to \hat{A}(X)$ induced by the mapping $x \mapsto x + \gamma(x)$, $x \in X$. Clearly, φ is injective and therefore, $\varphi\colon H \to \mathrm{gp}(\{a_x, x \in X\})$ is an isomorphism of groups thereby completing our proof.

Corollary. *Consider $\hat{L}$, as a group $\hat{L}^\circ$ under the operation introduced by the Campbell-Hausdorff formula* (see proposition 2.1), *and let $C = C(X)$ be its subgroup generated by X. Then C is a free group with set X of free generators.*

By the definition of the operation in $\hat{L}^\circ$, this group is isomorphic with $\mathrm{e}^{\hat{L}}$. Under this homomorphism C is isomorphic to the subgroup $T = \mathrm{gp}(\{\mathrm{e}^x, x \in X\})$ of $\mathrm{e}^{\hat{L}}$. But theorem 2, however, gives T being a free group freely generated by $\{\mathrm{e}^x, x \in X\}$, and the corollary follows.

6.4. Commutator calculus

Given a group G and its elements a, b, denote by $[\![a, b]\!]$ their *commutator* $a^{-1}b^{-1}ab$. Putting $a^{\pm b} = b^{-1}a^{\pm 1}b$, one may rewrite $[\![a, b]\!]$ in the form $[\![a, b]\!] = a^{-1}a^b$. In these terms one has the following identities holding in G

$$[\![ab, cd]\!] = [\![a, d]\!]^b\, [\![a, c]\!]^{db}\, [\![b, d]\!]\, [\![b, c]\!]^d \tag{1}$$

$$[\![[\![a, b]\!], c^a]\!]\, [\![[\![c, a]\!], b^c]\!]\, [\![[\![b, c]\!], a^b]\!] = 1\,. \tag{2}$$

To prove them, note that if $y \in G$ then $x \mapsto x^y$ is an automorphism of G. Then

$$[\![ab, cd]\!] = (ab)^{-1}\, (ab)^{cd} = b^{-1}a^{-1}a^d b b^{-1} a^{-d} a^{cd} b^{cd} =$$
$$= [\![a, d]\!]^b\, [\![a, c]\!]^{db}\, b^{-1} b^d b^{-d} b^{cd} = [\![a, d]\!]^b\, [\![a, c]\!]^{db}\, [\![b, d]\!]\, [\![b, c]\!]^d$$

and hence (1) is valid.

To prove (2), calculate directly the left hand side of this equality.

$$[\![[\![a, b]\!], c^a]\!] = [\![a, b]\!]^{-1}\, c^{-a} [\![a, b]\!]\, c^a = [\![b, a]\!]\, a^{-1}c^{-1}a [\![a, b]\!]\, a^{-1}ca =$$
$$= b^{-1}a^{-1}bc^{-1}b^{-1}aba^{-1}ca.$$

Permuting the letters cyclically, one gets

$$\llbracket\llbracket c, a\rrbracket, b^c\rrbracket = a^{-1}c^{-1}ab^{-1}a^{-1}cac^{-1}bc$$

and

$$\llbracket\llbracket b, c\rrbracket, a^b\rrbracket = c^{-1}b^{-1}ca^{-1}c^{-1}bcb^{-1}ab\ .$$

Multiplying these three factors and cancelling inverses one obtains unity.

Let further A, B be two subgroups of G. Denote by $\llbracket A, B\rrbracket$ the subgroup of G generated by $\llbracket a, b\rrbracket$, $a \in A$, $b \in B$. Since $\llbracket a, b\rrbracket^{-1} = \llbracket b, a\rrbracket$, clearly $\llbracket A, B\rrbracket = \llbracket B, A\rrbracket$. If A and B are normal in G then so is $\llbracket A, B\rrbracket$.

Lemma (PH. HALL). *Given a group G and three of its normal subgroups A, B, C. Then*

$$\llbracket\llbracket A, B\rrbracket, C\rrbracket \subseteq \llbracket\llbracket B, C\rrbracket, A\rrbracket\,\llbracket\llbracket C, A\rrbracket, B\rrbracket\ . \tag{3}$$

Proof. It suffices to prove that an element $u = \llbracket u_1, u_2\rrbracket$, $u_1 \in \llbracket A, B\rrbracket$, $u_2 \in C$ is an element of the right hand side of (3). Represent u_1 then as a product $u_1 = v_1v_2 \dots v_t$ of commutators $\llbracket x_i, y_i\rrbracket$, x_i, y_i being elements from A and B (or B and A). Using (1) has one

$$\llbracket v_1v_2 \dots v_t, u_2\rrbracket = \llbracket v_1, u_2\rrbracket^{v_2\dots v_t}\,\llbracket v_2, u_2\rrbracket^{v_3\dots v_t} \dots \llbracket v_t, u_2\rrbracket\ .$$

But $\llbracket v_i, u_2\rrbracket = \llbracket\llbracket x_i, y_i\rrbracket, (u_2^{x_i^{-1}})^{x_i}\rrbracket \overset{(2)}{=} \llbracket\llbracket y_i, u_2^{x_i^{-1}}\rrbracket, x_i^{y_i}\rrbracket^{-1}\,\llbracket\llbracket u_2^{x_i^{-1}}, x_i\rrbracket, y_i^{u_2^{x_i^{-1}}}\rrbracket^{-1} \subseteq$

$$\subseteq \llbracket\llbracket B, C\rrbracket, A\rrbracket\,\llbracket\llbracket C, A\rrbracket, B\rrbracket\ .$$

Noting that if X, Y are two normal subgroups of G, then $XY = YX$, we complete our proof.

Similarly to the Lie algebra case (cf. 1,6) one may introduce the *lower central*, the *upper central* and the *derived* series of a group. Then put $G = \gamma_1(G) = \delta_0(G)$ and define $\gamma_s(G) = \llbracket G, \gamma_{s-1}(G)\rrbracket$, $s > 1$; $\delta_r(G) = \llbracket\delta_{r-1}(G), \delta_{r-1}(G)\rrbracket$. Put also $\zeta_0(G) = \{1\}$ and define $\zeta_t(G)$ as a subgroup of G containing $\zeta_{t-1}(G)$ and such that $\zeta_t(G)/\zeta_{t-1}(G)$ is the centre of G. Call the series $\{\gamma_s(G)\}_{s\geq 1}$, $\{\delta_r(G)\}_{r\geq 0}$, and $\{\zeta_t(G)\}_{t\geq 0}$ respectively the *lower central*, the *derived* and the *upper central* series of G. Clearly, the two former of them are nonincreasing and the latter one is nondecreasing. Moreover, all the subsequent factors of these series, that is γ_i/γ_{i+1}, δ_i/δ_{i+1}, ζ_i/ζ_{i-1}, are abelian.

The lower central series of G is a partial case of so called *central filtrations* of a group G. Namely, a nonincreasing family $\{G_n, n = 1, 2, \dots\}$ of normal subgroups of G is called a *central filtration* of G if for all $s, r = 1, 2, \dots$, $\llbracket G_s, G_r\rrbracket \subseteq G_{s+r}$. For, by the lemma, $\llbracket\gamma_s(G), \gamma_r(G)\rrbracket = \llbracket\llbracket G, \gamma_{s-1}(G)\rrbracket, \gamma_r(G)\rrbracket \subseteq \llbracket\llbracket\gamma_{s-1}(G), \gamma_r(G)\rrbracket, G\rrbracket\,\llbracket\gamma_{s-1}(G), \gamma_{r+1}(G)\rrbracket$ and so the induction over s applies, the statement being trivial if $s = 1$.

Let now $f = \{G_n\}_{n\geq 1}$ be a central filtration of a group. Then $L_n(G) = G_n/G_{n+1}$ is an abelian group. Introduce an abelian group

$$L^f(G) = \bigoplus_{n\geq 1} L^f_n(G)$$

and turn it into a *graded Lie ring*, that is a graded Lie algebra over integers, defining the commutator of elements $\xi \in L^f_s$, $\eta \in L^f_r$ by the rule: if $\xi = xG_{s+1}$, $\eta = yG_{r+1}$, then $[\xi, \eta] = \llbracket x, y\rrbracket G_{s+r+1} \in L^f_{s+r}$. This definition does not depend on the choice of representatives in ξ and η, for if $x' = xz$, $y' = yw$, $z \in G_{s+1}$, $w \in G_{r+1}$, then by (1), $\llbracket x', y'\rrbracket = \llbracket x, w\rrbracket^z\,\llbracket x, y\rrbracket^{wz}\,\llbracket z, w\rrbracket\,\llbracket z, y\rrbracket^w \equiv \llbracket x, y\rrbracket \bmod G_{s+r+1}$ by the definition of filtrations. The Jacobi identity and the distributivity laws may be proved also only for the "homogeneous" elements and follow trivially from (2) and (1). Moreover, arguing as in the proof of the lemma, one sees that $n[\xi, \eta] = [n\xi, \eta] = [\xi, n\eta]$. (For, $n[\xi, \eta] = [\![x, y]\!]^n G_{s+r+1} = [\![x^n, y]\!] G_{s+r+1} = [\![x, y^n]\!] G_{r+s+1}$.) An example of such a Lie ring, constructed by the lower central series of a group was given in 1,2 (iv). A Lie ring obtained in this way will be denoted simply by $L(G)$ and called a *Lie ring of a group G*.

Our main goal in this section will be using of the Lie ring $L(F)$ of a free group F to describe the structure of the lower central series of this latter.

Let thus $F = F(X)$ be a free group with free generating set X, $A = A(X)$, $L = L(X)$, A, L, M be the algebras introduced in the previous sections. Identify F with a subgroup of $1 + M \subseteq \hat{A}$ using an isomorphism extending the mapping $x \mapsto 1 + x + \gamma(x)$, $\gamma(x) \in \hat{A}_{(2)}(X)$ where $x \in X$, M^n is the set of elements a from A having the form $a = \sum_{s \geqq n} a_s$, $a_s \in A^s$. Then we show that $\gamma_n(F) \subseteq 1 + M^n$. Indeed, $\gamma_1(F) = F \subseteq 1 + M$. If, further $c = [\![g, h]\!]$, $g \in \gamma_{n-1}(F)$, $h \in F$ then by the induction hypothesis, $g = 1 + g_{n-1} + g_n + \dots$, $h = 1 + h_1 + h_2 + \dots$.

An immediate calculation gives

$$c = g^{-1}h^{-1}gh =$$
$$= (1 - g_{n-1} + g_n' + \dots)(1 - h_1 + h_2' + \dots)(1 + g_{n-1} + g_n + \dots)(1 + h_1 + h_2 + \dots) =$$
$$= \left(1 + (g_{n-1}h - hg_{n-1}) + c_{n+1} + \dots\right), \quad c_{n+1} \in A_{n+1}. \tag{4}$$

Denote now by F_n the intersection of F and $1 + M^n$. Then we have

Proposition 1. *$\{F_n\}$, $n \geqq 1$, is a central filtration of F such that $F_n \supseteq \gamma_n(F)$. Consequently the intersection $\bigcap_{n=1}^{\infty} \gamma_n(F)$ of the members of the lower central series of a free group is trivial.*

Proof. Given $g = 1 + g_s + g_{s+1} + \dots \in F_s$ and $h = 1 + h_r + h_{r+1} + \dots \in F_r$, the same calculation as in (4) shows that

$$c = [\![g, h]\!] = 1 + (g_s h_r - h_r g_s) + c_{s+r+1} + \dots, \qquad c_{s+r+1} \in A_{s+r+1}, \tag{5}$$

that is $c \in F_{s+r}$. The inclusion $F_n \supseteq \gamma_n(F)$ is proved by (4). The last statement is trivial since $\bigcap_{n=1}^{\infty} \gamma_n(F) \subseteq \bigcap_{n=1}^{\infty} 1 + M^n = \{1\}$. Thus we are finished.

Now we can formulate the main

Theorem 3. *Let $L = L(X)$ be a free Lie ring, $F = F(X)$ be a free group, $L(F)$ be a Lie ring of F. Let ξ_x be an element of $L(F)$ having the form $x\gamma_2(F)$, $x \in X$. Then the mapping $x \mapsto \xi_x$, $x \in X$, extends to an isomorphism $\varphi: L \to L(F)$ of graded Lie rings.*

Proof. By the definition of a free Lie algebra such a homomorphism φ exists and we have to prove that it is bijective. Using proposition 1, we first construct a Lie ring $L^f(F)$ using the above introduced filtration $\{F_n\}_{n \geqq 1}$. Since $F_n \supseteq \gamma_n(F) \supseteq F_{n+1} \supseteq \gamma_{n+1}(F)$, there exists a natural homomorphism $L_n(F) \to L_n^f(F)$ of abelian groups which obviously extends to a homomorphism ψ of Lie rings $L(F)$ into $L^f(F)$. One more mapping to be considered is that of $L^f(F)$ into A. Let $g = 1 + g_s + g_{s+1} + \dots \in F_s$ and $\mu = gF_{s+1}$. Put $\chi(\mu) = g_s$. The definition is easily seen to be independent of the choice of representatives and (5) shows that χ is a homomorphism of Lie rings. Consider now the product $\tau = \chi\psi\varphi: L \to A$ of homomorphisms of Lie rings. By our construction $\tau(x) = \chi\psi\varphi(x) = \chi\psi(\xi_x) = \chi\left((1 + x + \gamma(x))\, F_2\right) = x$. By the theorem 2,2.1 applied to a Z-subalgebra of A generated by X, τ is injective. The same is hence true for φ. Moreover, it is easily seen from the definition of a Lie ring of a group that $L(F)$ is generated by $\{\xi_x, x \in X\}$. This shows φ bijective and the trivial fact $\varphi(L_n) \subseteq L_n(F)$ completes the proof of the theorem.

Corollary. (i) *$\gamma_n(F)/\gamma_{n+1}(F)$ is a free abelian group canonically isomorphic with $L_n(X)$, $n = 1, 2, \ldots$.*

(ii) $\gamma_n(F) = F_n$, $n = 1, 2, \ldots$.

Proof. (i) follows immediately from theorem 3 and theorem 2,2.1.

(ii) Clearly $\gamma_1(F) = F_1 = F$. Having assumed $\gamma_{n-1}(F) = F_{n-1}$, remark that the mapping $L_n(F) \to L_n^f(F)$ is injective. But by the definition now its kernel equals $F_n/\gamma_n(F)$ which proves $F_n = \gamma_n(F)$ and the whole corollary.

6.5. Nilpotent groups

A group G is called *nilpotent of class* c if, for some $c \geqq 1$, $\gamma_c(G) \neq \{1\}$ and $\gamma_{c+1}(G) = \{1\}$. For example the group $UT(n, k)$ of all unitriangular matrices over some ring k is nilpotent of class $n - 1$. On the contrary, the group $T(n, k)$ of all nondegenerate upper triangular matrices over, say, the rational numbers is not nilpotent, for as the reader may easily check up, if

$$A = \begin{pmatrix} 1 & 0 & \ldots & 0 & 1 \\ 0 & 1 & \ldots & 0 & 0 \\ \cdot & \cdot & \cdot & \cdot & \cdot \\ 0 & 0 & \ldots & 1 & 0 \\ 0 & 0 & \ldots & 0 & 1 \end{pmatrix}, \qquad B = \begin{pmatrix} 1 & 0 & \ldots & 0 & 0 \\ 0 & 1 & \ldots & 0 & 0 \\ \cdot & \cdot & \cdot & \cdot & \cdot \\ 0 & 0 & \ldots & 1 & 0 \\ 0 & 0 & \ldots & 0 & -1 \end{pmatrix}$$

then $[\![A, B]\!] = A^{-2}$, and hence $A^{\pm 2^n} \in \gamma_n(T)$ for any $n \geqq 1$. Applying the arguments of section 1,7 one may easily prove a group G to have nilpotency class c if and only if c is the least positive integer such that for any $g_1, g_2, \ldots, g_{c+1}$ and any arrangement of brackets

$$[\![g_1, g_2, \ldots, g_{c+1}]\!] = 1\,,$$

or, equivalently, $[\![x_1, x_2, \ldots, x_{c+1}]\!] = [\![x_1, [\![x_2, [\![\ldots [\![x_c, x_{c+1}]\!] \ldots]\!]]\!]]\!] \equiv 1$ is an *identity* in G. As in 1,7, this yields that any subalgebra, quotient algebra and cartesian product of nilpotent groups of some class $\leqq c$ is itself nilpotent of class $\leqq c$. (In other words, the class of all nilpotent groups of nilpotency class $\leqq c$ is a "variety", denoted by $\mathfrak{N}_c$.)

For our future purposes we need some more detailed study of finitely generated torsion free nilpotent algebras. We start with

Proposition 1. *Let G be a nilpotent group, generated by a finite system $\{g_1, g_2, \ldots, g_m\}$ of its elements. Then, for any $s \geqq 1$, $\gamma_s(G)$ may be generated by the finite set of all nontrivial commutators of the form*

$$[\![g_{i_1}^{\varepsilon_1}, g_{i_2}^{\varepsilon_2}, \ldots, g_{i_r}^{\varepsilon_r}]\!], \qquad 1 \leqq i_1, \ldots, i_r \leqq m; \qquad \varepsilon_i = \pm 1\,, \tag{1}$$
$$i = 1, 2, \ldots, r, \qquad r \geqq s\,.$$

Moreover, any subgroup A of G is finitely generated.

Proof. Indeed, it suffices to prove that any commutator w of the form $[\![u, v]\!] \in \gamma_{n+1}(G)$, where $u \in \gamma_n(G)$, $v \in G$, can be expressed through (1) and their inverses. Thus let $u = u_1 u_2 \ldots u_t$, u_i have the form (1) with $r \geqq n$, and $v = v_1 v_2 \ldots v_q$, v_i being the generators g_j or their inverses. Using 4 (1), w may be represented as the product of the commutators having the form (1) with $r \geqq n$, and their conjugates by some elements from G. But note, that $a^b = a[\![a, b]\!]$ and

$$a^{b_1 b_2} = (a^{b_1})^{b_2} = a^{b_1}[\![a, b_1]\!]^{b_2} = a[\![a, b_2]\!]\,[\![a, b_1]\!]\,[\![[\![a, b_1]\!], b_2]\!]\,.$$

These relations enable us to represent w in the desired form and hence to prove the first part of the proposition.

The second part will be proved using induction over c, the statement being trivial for $c = 0, 1$. Thus let $c > 1$ and A be a subgroup of G. Then $A\gamma_c(G)/\gamma_c(G)$ is a finitely generated subgroup of $G/\gamma_c(G)$, whose nilpotency class is less than that of G. Let $a_1, a_2, \ldots, a_q \in A$ be such that $a_1\gamma_c, \ldots, a_q\gamma_c$ generate $A\gamma_c/\gamma_c$ and $a_{q+1}, \ldots, a_t$ be the generators of the subgroup $A \cap \gamma_c$ of a *finitely generated abelian group* $\gamma_c = \gamma_c(G)$ (this is true by the first part of the proof). Then, evidently, given $w \in A$, one has $w\gamma_c = w(a_1\gamma_c, \ldots, a_q\gamma_c) = w(a_1, \ldots, a_q)\,\gamma_c$ and $w = w(a_1, \ldots, a_q)w'$, where $w' \in \gamma_c$ and therefore $w' \in \gamma_c \cap A$, whence $w' = w'(a_{q+1}, \ldots, a_t)$. This completes the proof of the proposition.

Corollary. *Any nondecreasing chain of subgroups of a finitely generated nilpotent group breaks at some finite step, that is if $H_1 \subseteq H_2 \subseteq \ldots \subseteq H_q \subseteq H_{q+1} \subseteq \ldots$ is such a chain, then for some $q \geqq 1$, $H_q = H_{q+1} = \ldots$.*

As in 4,8, call a group G *monolithic* if the intersection M of all its nontrivial normal subgroups is nontrivial. The least nontrivial normal subgroup M is called a *monolith* of G.

Theorem 4. *Every finitely generated monolithic nilpotent group is a finite p-group for some prime p.*

First we state without proof a simple lemma whose Lie algebra versions are proved in 1,7.

Lemma 1. (i) *In a nilpotent group G the lower and the upper central series (see section 4) have the same length equal to the nilpotency class of G.*

(ii) *Let H be a subgroup of a nilpotent group G and $N_G(H)$ be its normalizer, that is the subgroup formed by $x \in G$ such that $x^{-1}Hx \subseteq H$, then $H \neq G$ implies $N_G(H) \neq H$.*

(iii) *Let H be a subgroup of G such that $H\gamma_2(G) = G$. Then $H = G$.*

Lemma 2. *A finitely generated nilpotent group G with finite centre is finite.*

Proof. Induction by the nilpotency class c of G. If $c = 1$ then $G = \zeta_1(G)$, the centre of G, and there is nothing to prove. Otherwise, we consider $G_1 = G/\zeta_1(G)$ and prove that $\zeta_2(G)/\zeta_1(G)$, the centre of G_1 is finite. Then the induction hypothesis applies. But $\zeta_2(G)/\zeta_1(G)$ is a finitely generated abelian group and it suffices to show that all its elements have finite order. To this end let $n = |\zeta_1(G)|$ and x be any element from $\zeta_2(G)$. We prove that $x^n \in \zeta_1(G)$. For, $[\![x^n, g]\!]$ by (3) may be rewritten in the form

$$[\![x^n, g]\!] = [\![x, g]\!]^{x^{n-1}}\,[\![x, g]\!]^{x^{n-2}} \ldots [\![x, g]\!] = [\![x, g]\!]^n = 1\,,$$

since $[\![x, g]\!]$ is central and $|\zeta_1(G)| = n$. This completes the proof of the lemma.

Lemma 3. *Every finite monolithic nilpotent group G is a p-group for some prime p.*

Proof. Suppose not. Then there exist two primes p, q such that S_p, S_q, the Sylow p- and q-subgroups, are nontrivial and hence different from G. By a theorem of Sylow if N_p is a normalizer of say S_p in G, then $N_G(N_p) = N_p$. This implies by the lemma 1 (ii), that $N_p = G$ and hence that S_p is a normal subgroup of G. The same is true of course for S_q. But this leads to the contradiction for, on one hand, $S_p \cap S_q = \{1\}$ and, on the other hand, $S_p \cap S_q = M \neq \{1\}$ since G is monolithic. Hence we are finished.

Proof of the theorem. The assertion is trivial if $c = 0$. Thus assume $\zeta_1(G) \neq \{1\}$ and consider a monolith M of G. Since any subgroup of $\zeta_1(G)$ is normal in G and using a decomposition of a finitely generated abelian group into a direct product of infinite or prime power order cyclic groups, one observes that $\zeta_1(G)$ is cyclic and M is a minimal nontrivial subgroup of $\zeta_1(G)$. This yields clearly that $\zeta_1(G)$ is a cyclic group of order p^t, p being some prime. It remains then to apply lemmas 2 and 3.

An important corollary of the lemma 2 is

Corollary. *Any finitely generated torsion free nilpotent group is finite.*

6.6. Malcev completion

Define one more class of groups termed as *D-groups*. By a *D-group* we shall mean a group G such that for each integer $n \neq 0$ and each $g \in G$ the equation $x^n = g$ is *uniquely* solvable in G. The role of D-groups for our purposes may be explained by the fact that given a complete topological Lie algebra B over a field k of characteristic 0, its group B° (see proposition 2.1) under Campbell-Hausdorff multiplication $\circ$ is a D-group. For x^n then equals nx and therefore $\frac{1}{n} g$ is a single solution of $x^n = g$, $n \neq 0$.

Our main goal in this section is to prove that any nilpotent torsion free group of class c is a subgroup of a nilpotent D-group of the same class. To formulate this result more correctly we need the concept of an *isolated* subgroup of a group. Namely, a subgroup H of a group G is called *isolated* if, given $g \in G$ such that $g^n \in H$ for some $n \neq 0$, then $g \in H$. The least isolated subgroup $I_G(H)$ of G containing H is called an *isolator* of H in G. Denote also by $\sqrt{H}$ a set of elements $g \in G$ such that for some nonzero integer n, $g^n \in \sqrt{H}$. Evidently, if $\sqrt{H}$ is a subgroup of G, then $I_G(H) = \sqrt{H}$. The next lemma shows that this is so in the case of nilpotent groups.

Lemma 1. *Let H be a subgroup of a nilpotent group G. Then $I_G(H) = \sqrt{H}$.*

Proof. As it is remarked above it suffices to prove that H is a subgroup of G. We proceed by induction on the nilpotency class c of G, the case $c = 1$ being trivial. Thus assume $c > 1$ and x^n, $y^m \in H$. Put $A = \mathrm{gp}(\{x, y\})$ and consider $\{\gamma_i(A)\}_{1 \le i \le c}$, the lower central series of A. Apply the induction hypothesis to the subgroup $\overline{A \cap H} = (A \cap H)\, \gamma_c(A)/\gamma_c(A)$ as a subgroup of $A/\gamma_c(A)$. Then for each $\bar{a} = a\gamma_c(A)$ there exists an integer $s \neq 0$ such that $\bar{a}^s \in \overline{A \cap H}$, or that $a^s \in (A \cap H)\, \gamma_c(A)$. We prove further that $\gamma_c(A) \subseteq \sqrt{A \cap H}$. Let first $z \in \gamma_c(A)$ have the form $[\![a, b]\!]$, $a \in \gamma_{c-1}(A)$, $b \in A$, and $a^s \in$

$\in (A \cap H)\,\gamma_c(A)$, $b^r \in (A \cap H)\,\gamma_c(A)$, that is $a^s = h_1 z_1$, $b^r = h_2 z_2$. Using formula 4 (1) and the centrality of all commutators arising in its application, one sees $[\![a, b]\!]^{sr} = [\![a^s, b^r]\!] = [\![h_1 z_1, h_2 z_2]\!] = [\![h_1, h_2]\!] \in A \cap H$ and hence $[\![a, b]\!] \in A \cap H$. An arbitrary element of $\gamma_c(A)$ is a product of pairwise commuting commutators having the form of the above considered one. This evidently proves our statement about $\gamma_c(A)$. Remember now that we have found, for any $a \in A$, a nonzero integer s with $a^s = hz$, $h \in A \cap H$, $z \in \gamma_c(A)$. By the just proved statement, however, for some nonzero r, $z^r \in A \cap H$. Since h, z commute, $a^{sr} = h^r z^r \in A \cap H$ and the lemma follows.

One more example of isolated subgroups is given by the next lemma including also an important auxiliary result on the uniqueness of the roots in a torsion free nilpotent group.

Lemma 2. *Given a torsion free nilpotent group G.*

(i) *If $x^n = y^n$ for some $x, y \in G$ and a nonzero integer n, then $x = y$.*

(ii) *The members of the upper central series of G are isolated.*

Moreover $I_G(\gamma_s(G)) = \sqrt{\gamma_s(G)}$ in any group G.

Proof. (i) We proceed using induction over c, the nilpotency class of G, the assertion being easy for the case $c = 1$. Consider $G_1 = \mathrm{gp}(\{x\}, \gamma_2(G))$. Then obviously $\gamma_c(G_1) = \{1\}$ and $u = y^{-n} x y^n = x[\![x, y^n]\!] \in G_1$. Now $x^n = y^n$ gives $u^n = y^{-n} x^n y^n = x^n$ whence by the induction hypothesis $x = u = y^{-n} x y^n$ and $y^n = x^{-1} y^n x$. Similarly, $y = x^{-1} y x$, hence $x^n = y^n$ implies $(xy^{-1})^n = 1$ and $x = y$ by our conditions.

(ii) It suffices to limit ourselves to the case of $\zeta_1(G)$, the remaining being trivial. So far, let $g^n \in \zeta_1(G)$ for some $g \in G$ and some nonzero n. Then for any $a \in G$, $g^n = a^{-1} g^n a = (a^{-1} g a)^n$ and therefore by (i) $g = a^{-1} g a$ which yields $g \in \zeta_1(G)$. The last assertion follows from lemma 1 by easy argument modulo $\gamma_s(G)$ and completes the proof of the lemma.

An interesting way to obtain D-groups is shown by

Lemma 3. *Let π be an infinite set of primes and, for each $p \in \pi$, let G_p be a nontrivial finite p-group of a nilpotency class $\leqq c$ for some fixed $c \geqq 1$. Let $T = C\prod_{p \in \pi} G_p$ be the cartesian product of all G_p, $p \in \pi$, and $S = \prod_{p \in \pi} G_p$ be its subgroup equal to the direct product of the same set of groups. Then T/S is a D-group.*

Proof. Note first that in any group the equation $x^n = a$ with $a \in G$ and a nonzero n coprime to the order of G is uniquely solvable. Remark also that G is a D-group if and only if any equation of the form $x^q = a$, $a \in G$, q being any prime, is uniquely solvable in G.

Using these two remarks one sees that it suffices that the equation $x^p = a$ were uniquely solvable in T/S, where $p \in \pi$. Let then $a = fS$, f being some function from T. Then multiplying by a suitable element from S one may think that $f(p) = 1$. By the initial remark of the proof for some $b \in T$ one has $b^p = f$ and hence bS is the solution of $x^p = a$ in T/S.

To prove the uniqueness of the solution remark that by our conditions T has nilpotency class $\leqq c$ and that $S = \sqrt{\{1\}}$ is an isolated subgroup of T. Hence T/S is a torsion free nilpotent group so that lemma 2 (i) applies and we are finished.

And the last lemma we need to be able to embed any torsion free nilpotent group in a nilpotent D-group is

Lemma 4. *Let G be a finitely generated torsion free nilpotent group, p be a prime, G^{p^s} be a subgroup generated by g^{p^s} for all $g \in G$, $s = 0, 1, 2, \ldots$. Then $\bigcap_{s=0}^{\infty} G^{p^s} = \{1\}$.*

Proof. We show that for any $1 \neq x \in G$ there exists an $s > 0$ such that $x \notin G^{p^s}$. Use induction by the nilpotency class of G. Then by lemma 2 (ii) the induction hypothesis applies to $G/\zeta_1(G)$. Thus if $x \notin \zeta_1(G)$ then $x\zeta_1(G) \notin (G/\zeta_1(G))^{p^s} = G^{p^s}\zeta_1(G)/\zeta_1(G)$ for some $s \geqq 1$. Clearly then $x \notin G^{p^s}$ and all is fine. Thus let $1 \neq x \in \zeta_1(G)$. Consider a nonempty nondecreasing chain of normal subgroups $K_i \lhd G$, $i = 1, 2, \ldots$, such that $x \notin K_i$ and $x^p \in K_i$ (such a chain exists since $\{x^p\}$, the cyclic subgroup generated by x^p, enjoys these two properties). By the corollary of proposition 5.1 this chain possesses a maximal element K. Now G/K by the maximality of K is a monolithic group for any of its normal subgroups contains $\{1\} \neq M = \text{gp}(\{x\})K/K$. By the theorem 5.3 then G/K is a finite q-group for some prime q. But G/K contains an element xK of order p. Consequently G/K is in fact a finite p-group of order, say, p^l, $l \geqq 1$. Clearly this yields $K \supseteq G^{p^l}$ and $x \notin G^{p^l}$. The proof is now complete.

Theorem 5 (Malcev). *Every torsion free nilpotent group G of nilpotency class c can be embedded in a nilpotent D-group G^* of the same nilpotency class and such that any homomorphism φ of G into any D-group Q uniquely extends to a homomorphism φ^* of G^* into Q.*

Proof. Assume first G finitely generated.

Since then G is countable there exists an injection $x \mapsto p_x$ of G onto some infinite set π of primes. Form $T = C\prod_{p\in\pi} G_p$ and $S = \prod_{p\in\pi} G_p$, where G_p for $p = p_x$ is a quotient group G/G^{p^s}, s being an integer such that $x \notin G^{p^s}$ (see the preceding lemma 4). By the final corollary of the section 5 each G_p is a finite p-group. Now by lemma 3, $P = T/S$ is a D-group. Define $\iota: G \to P$ by $\iota(y) = f_yS$, where $f_y(p)$ is an image of y under the natural epimorphism of G onto G_p. By the definition ι is a homomorphism. Assume next $\iota(y) = 1$ for some $y \in G$. Then $f_y \in S$ so that $f_y(p) \neq 1$ only for some finite set of primes p. But clearly then there exists a nonzero integer n such that $f_{y^n}(p) = 1$ for all $p \in \pi$. By lemma 4 this yields $y^n = 1$. So ι is an injective homomorphism, and using it we may identify G with its image in P.

Take now for G^* the isolator $I_p(G)$ in P. Then G^* is a D-group of nilpotency class $\leqq c$ and thus, by the lemma 1, $G^* = \sqrt{G}$. Denote by $g^{1/n}$ the unique element $x \in P$ such that $x^n = g \in P$. Then any element from G^* has the form $g^{1/n}$, $g \in G$ and $n \neq 0$. Let now Q be a D-group and $\varphi: G \to Q$ be a homomorphism of G into Q. Define then $\varphi^*: G^* \to Q$ by putting $\varphi^*(g^{1/n}) = (\varphi(g))^{1/n}$. The definition is unambiguous for $g^{1/n} = h^{1/m}$ implies $(\varphi^*(g^{1/n}))^{mn} = (\varphi(g^{1/n}))^{mn} = \varphi(g)^m = \varphi(h)^n = (\varphi^*(h^{1/n}))^{mn}$ since $g^m = h^n$. As, further, Q is a D-group, $\varphi^*(g^{1/n}) = \varphi^*(h^{1/m})$. Similarly one may verify φ^* to be a homomorphism, and so the case of a finitely generated group is exhausted.

To deal with the general case, given a torsion free nilpotent group G, consider the set of all symbols $g^{1/n}$, $g \in G$, n being a positive integer, and say that $g^{1/n}$ is *equivalent* to $h^{1/m}$ (denoted by $g^{1/n} \sim h^{1/m}$) if $g^m = h^n$. Let G^* be the set of the classes of the symbols formed by this equivalence. Define the multiplication in G^* by the rule $g_1^{1/m}g_2^{1/l} = g_3^{1/n}$ if this equality holds in G_1^*, $G_1 = \text{gp}(\{g_1, g_2\}) \subseteq G$. G_1^* exists by the first part of the proof. Assume that $g_1^{1/m} \sim \bar{g}_1^{1/m'}$, $g_2^{1/l} \sim \bar{g}_2^{1/l'}$ and let $\bar{g}_1^{1/m'}\bar{g}_2^{1/l'} = \bar{g}_3^{1/n'}$. Consider $G_2 = \text{gp}(\{g_1, g_2, \bar{g}_1, \bar{g}_2\}) \subseteq G$ and $G_2^* = \sqrt{G}$. Then $(g_1^{1/m})^{mm'} = g_1^{m'} = \bar{g}_1^m = (\bar{g}_1^{1/m'})^{mm'}$ so that $g_1^{1/m} = \bar{g}_1^{1/m'}$ in G_2^*.

Similarly $g_2^{1/l} = \bar{g}_2^{1/l'}$ in G_2^*. But then $g_3^{1/n} = \bar{g}_3^{1/n'}$ and therefore $g_3^{n'} = \bar{g}_3^n$. This yields $g_3^{1/n} \sim \bar{g}_3^{1/n'}$ so that the multiplication in G is unambiguous. The other properties required may be checked up analogously.

Thus the proof of the theorem is complete.

The unique D-group G^* constructed in theorem 4 is called the *Malcev completion* of the torsion free nilpotent group G.

6.7. Isomorphism of categories of nilpotent Lie Q-algebras and nilpotent D-groups

Let S be a nilpotent Lie algebra over some field k of characteristic 0, and $\mathcal{F}$ be a filter of ideals of S consisting of a single element $\{0\}$. Then S becomes a complete topological algebra with discrete topology. As before (see proposition 2.1) denote by S° the group of S obtained by the Campbell-Hausdorff formula.

Using the remark from section 2 and the discreteness of the topology in S, we see that, given any ideal $R \lhd S$, R° is a normal subgroup of S° and $S^\circ/R^\circ \cong (S/R)^\circ$.

Lemma 1. $\gamma_n(S^\circ) = (S^n)^\circ$, $n = 1, 2, \ldots$.

Proof. Let c be the least number with $S^{c+1} = \{0\}$. Direct calculations show (as in proposition 4.1)

$$[\![a_1, a_2, \ldots, a_n]\!] = [a_1, a_2, \ldots, a_n] + t, \quad t \in S^{n+1}, \quad n = 1, 2, \ldots, \tag{1}$$

the arrangement of brackets in both sides of (1) is the same. From this equality one immediately has $(S^c)^\circ = \gamma_c(S^\circ)$. Using induction by the nilpotency class of S, one has $\gamma_r(S^\circ)/\gamma_c(S^\circ) = \gamma_r(S^\circ/\gamma_c(S^\circ)) = \gamma_r((S/S^c)^\circ) = (S^r/S^c)^\circ = (S^r)^\circ/(S^c)^\circ = (S^r)^\circ/\gamma_c(S^\circ)$, whence $\gamma_r(S^\circ) = (S^r)^\circ$, $r = 1, 2, \ldots, c$, which proves the lemma.

Lemma 2. *Any isolated subgroup of S° has the form R°, R being a subalgebra of S; besides if R° is a normal subgroup of S° then R is an ideal of S.*

Proof. Denote this subgroup by T and consider a subalgebra R generated by T. Then the explicit form of the Campbell-Hausdorff formula shows that $T + R^2$ is a subalgebra of S, for $x + y = x \circ y + z$, $z \in R^2$, and $[x, y] \in R^2$, where $x, y \in T$. Then $R = T + R^2$ and $R^\circ/\gamma_2(R^\circ) = R/(T^2)^\circ = (R/R^2)^\circ$, that is R° is generated modulo $\gamma_2(R^\circ)$ by T. Using lemma 5,1 (iii), $R^\circ = T$.

Assume now R° to be normal in S° and choose $x \in S$, $y \in R$. Let $P = \operatorname{alg}(\{x, y\})$. Then $R^\circ \cap P^\circ$ is a normal subgroup of P° with abelian $P^\circ/R^\circ \cap P^\circ$. This yields $R^\circ \cap P^\circ \supseteq \gamma_2(P^\circ) = P^2$ and $[x, y] \in R \cap P \subseteq R$, that is $R \lhd S$ and we are finished.

We need for the sequel two new concepts. Namely, given a nonempty set X and a positive integer c, consider a factor group $G = F/\gamma_{c+1}(F)$, where $F = F(X)$ is a free group in X. Then clearly given any map of X into a group having nilpotency class $\leqq c$ there exists only one homomorphism of G into the same group induced by this map. The group G is therefore called a *free nilpotent group* of nilpotency class c. As is it seen from the final corollary of section 4, G is torsion free and thus possesses a Malcev completion G^* called a *free nilpotent D-group* of nilpotency class c. We identify X with its image in G and G^*.

Let S be a free nilpotent Lie algebra of nilpotency class c with the same set X of free generators.

Lemma 3. $S^\circ \cong G^*$.

Proof. Using notations and results of section 4, the subgroup $C = \mathrm{gp}(\{e^x, x \in X\})$ of $1 + M^n \subseteq \hat{A}$ generated by e^x for all $x \in X$ is canonically isomorphic with $F(X) = F$ and $\gamma_n(C) = C \cap (1 + M^n)$ for all $n \geqq 1$. But then $D = \mathrm{gp}(X) \cong F(X)$ and $\gamma_n(D) = D \cap (\hat{L}^n)^\circ$. By the isomorphism theorem and the remark from section 2,

$$D/\gamma_n(D) \cong D^\circ(\hat{L}^n)^\circ/(\hat{L}^n)^\circ \subseteq \hat{L}^\circ/(\hat{L}^n)^\circ \cong (\hat{L}/\hat{L}^n)^\circ = S^\circ. \quad (1)$$

To complete the proof it now suffices to show that S° is an isolator of the subgroup T of S° generated by X and by (1) isomorphic to $G \cong D/\gamma_n(D)$. Using lemma 1, $\gamma_2(S^\circ) = (S^2)^\circ$, and therefore S° is generated modulo $\gamma_2(S^\circ)$ by the $\frac{m_1}{m_2}x = x^{m_1/m_2}$, $x \in X$, $m_1 \in \mathbf{Z}$, that is $S^\circ = \sqrt{T}\,\gamma_2(S)$. Since always $\sqrt{T} \subseteq I_{S^\circ}(T)$, one has $S^\circ = I_{S^\circ}(T)\,\gamma_2(S)$ and by the already quoted lemma 5.1 (ii) $S^\circ = I_{S^\circ}(T)$. By the uniqueness of the completion now $S^\circ \cong G^*$.

Now we can prove the main theorem of this section.

Theorem 6. *Every nilpotent D-group is a group of some rational Lie algebra unique up to isomorphism.*

Proof. Given any nilpotent D-group B there exists a free nilpotent D-group $G^* = S^\circ$, for some free nilpotent Lie algebra S and its isolated normal subgroup R° such that $B \cong S^\circ/R^\circ \cong (S/R)^\circ$. This proves the surjectivity of the map $T \to T^\circ$, T being a nilpotent Lie algebra over $\mathbf{Q}$.

Assume then that $f\colon T^\circ \to U^\circ$ is a homomorphism of groups, U being another nilpotent Lie algebra over $\mathbf{Q}$. Represent T° in the form $T^\circ \cong S^\circ/R^\circ$ and let $\varepsilon\colon S \to T$ be a natural epimorphism of S onto T. Then $\varepsilon^\circ\colon S^\circ \to T^\circ$ (by the remark (ii) from the section 2) is a natural epimorphism of groups. Let now $u_x = (f\varepsilon^\circ)(x)$, $x \in X$, and consider a unique homomorphism $\pi\colon S \to U$ extending the map $x \mapsto u_x$ for all $x \in X$. Then π° is the unique homomorphism of $S^\circ \to U^\circ$ extending the same mapping so that $\pi^\circ = f\varepsilon^\circ$. This enables to define correctly a homomorphism of Lie algebras $\varphi\colon T \to U$ by the formula $\varphi(t) = \pi(\varepsilon^{-1}(t))$. Clearly then $\varphi^\circ = f$, the uniqueness of φ with $\varphi^\circ = f$ being evident. Now $T^\circ \cong U^\circ$ implies $T \cong U$ and the proof of the theo rem is complete.

Corollary 1. *The categories $\mathfrak{N}_\mathfrak{L}$ and $\mathfrak{N}_\mathfrak{D}$ of, respectively, nilpotent Lie algebras over $\mathbf{Q}$ and nilpotent D-groups are isomorphic.*

Proof. For, there exist by the theorem one-to-one correspondences $T \mapsto T^\circ$ of their objects and $\varphi \mapsto \varphi^\circ$ of their morphisms such that $(1_T)^\circ = 1_{T^\circ}$ and $(\varphi\psi)^\circ = \varphi^\circ\psi^\circ$.

Corollary 2. *Any torsion free nilpotent group G is a subgroup of a group S° of some nilpotent Lie algebra S over $\mathbf{Q}$ such that $\sqrt{G} = S^\circ$.*

Corollary 3. $I_G(\gamma_n(G)) = \gamma_n(G^*) \cap G$.

For $S = \mathbf{Q}G$ and thus $\gamma_n(G^*) = S^n = \mathbf{Q}\gamma_n(G)$, $n = 1, 2, \ldots$.

The final corollary of this section deals with the so called "residually torsion free nilpotent" groups. Namely, a group G is called *residually torsion free nilpotent* if, for any

$g \in G$, $g \neq 1$, there exists a normal subgroup N_g with $g \notin N_g$ such that G/N_g is torsion free nilpotent. This condition is equivalent to the following one. Let $I_n(G) = I_G(\gamma_n(G))$, $n = 1, 2, \ldots$, be the isolators of the members of the lower central series of G. Then the equivalent condition to the residual nilpotency is $\bigcap_{n=1}^{\infty} I_n(G) = \{1\}$. We leave the proof to the reader. By the reasons explained in the corollary 4 below the residually torsion free nilpotent groups will be termed as *Lie type* groups.

Corollary 4. *Any Lie type group is isomorphic to a subgroup of a group of some Lie algebra over* $\mathbb{Q}$.

Proof. Let $\{G^{(n)} = G/I_n, \varphi_{nm}: G^{(n)} \to G^{(m)}\}_{n \geq m \geq 1}$ be a projective system of groups constructed by G in the same way as in section 1 and $T^{(n)}$ be a Lie algebra such that $(T^{(n)})^\circ = (G^{(n)})^*$, $n = 1, 2, \ldots$. Denote also by ψ_{nm} epimorphisms $\psi_{nm}: T^{(n)} \to T^{(m)}$ such that $\psi_{nm}^\circ = \varphi_{nm}^*$. Put $\tilde{T} = \varprojlim T^{(n)}$ and turn T into a topological algebra in the same way as it is done in section 1. Now $G \subseteq \tilde{G} = \varprojlim G^{(n)} \subseteq \varprojlim (T^{(n)})^\circ = (\varprojlim T^{(n)})^\circ = \tilde{T}^\circ$ and the proof is complete.

6.8. Relatively free groups

All the definitions and simple results related to varieties of Lie algebras given in 3,1 may be easily adapted to the group case. Thus, a *variety of groups* $\mathfrak{V}$ is a class of groups satisfying some fixed system $\{v(x_1, x_2, \ldots, x_n) \equiv 1, v \in V \subseteq F(X)\}$ of *identical relations*, or simply identities. A nonempty class $\mathfrak{V}$ of groups is a variety if and only if $\mathfrak{V}$ is closed under taking homomorphic images, subgroups and cartesian products of its groups (cf. theorem 3,1.1). The verbal subgroup $V(G)$ of a group G corresponding to the set V of the *words* $v \in F$ is the (normal) *subgroup* generated by the values $v(g_1, g_2, \ldots, g_n)$, $g_1, g_2, \ldots, g_n \in G$. The complete set of the left hand sides of the identities holding in a variety $\mathfrak{V}$ is a verbal subgroup of a free group F_∞ of countable rank. Given a set $V \subseteq F_\infty$ and a set $X \neq \emptyset$, the group $F(X, \mathfrak{V}) = F(X)/V(F(X))$ is called a *free group of a variety* $\mathfrak{V}$ determined by V, or a *relatively free group* of a variety $\mathfrak{V}$, or simply a $\mathfrak{V}$*-free group* with set X of free generators. These are the groups whose structure will interest us in the remaining part of this chapter.

Call a variety of groups a *Lie type* one if its free groups are of Lie type. Thus the variety $\mathfrak{O}$ of all groups and the varieties $\mathfrak{N}_c$, $c \geq 1$, of all nilpotent groups of nilpotency class $\leq c$, etc., are Lie type ones. Lie type groups being closely connected with Lie algebras, we prove that the same is true for Lie type varieties of groups and Lie algebra varieties.

But first some auxiliary results on isolated central series of a group.

Proposition 1. *Given an arbitrary group* G, *denote by* I_n, $n \geq 1$, *the subgroups* $I_n(G) = I_G(\gamma_n(G))$. *Then*

(i) *all* I_n *are fully invariant, i. e. endomorphism closed;*

(ii) *the set* $\{I_n\}_{n \geq 1}$ *is a central filtration of* G;

(iii) *if* G *is finitely generated then* $\mathrm{rk}(I_n/I_{n+1}) = \mathrm{rk}(\gamma_n/\gamma_{n+1})$.

(By the *rank* rk of a finitely generated abelian group we mean the number of infinite cyclic groups in its representation as a direct product of cyclic groups.)

Proof. By 6.2 lemma $I_s = \sqrt{\gamma_s}$. This implies the fully invariance of I_s in G for all $s = 1, 2, \ldots$ and proves (i). We prove now $[\![I_r, I_s]\!] \subseteq I_{r+s}$. Indeed, take $x \in I_r$, $y \in I_s$. Then for some $n, m \neq 0$, $x^n \in \gamma_r$, $y^m \in \gamma_s$ and therefore, by the results of section 4, $[\![x^n, y^m]\!] \in \gamma_{r+s} \subseteq I_{r+s}$. Let $\tilde{x}, \tilde{y}$ be images of x, y in a torsion free nilpotent group $\tilde{G} = G/I_{s+r}$. Then one has $\tilde{x}^n = \tilde{y}^{-m}\tilde{x}^n\tilde{y}^m = (\tilde{y}^{-m}\tilde{x}\tilde{y}^m)^n$. Again 6.2 implies $\tilde{y} = \tilde{x}\tilde{y}\tilde{x}^{-1}$ or $[\![\tilde{x}, \tilde{y}]\!] = 1$. Finally, $[\![x, y]\!] \in I_{r+s}$ and (ii) is complete. To prove (iii) recall that if A is a free abelian group of some finite rank and B is its subgroup of finite index then $\mathrm{rk}(A) = \mathrm{rk}(B)$. Now $\gamma_n \cap I_{n+1}/\gamma_{n+1}$ is a finite group and $\gamma_n/\gamma_n \cap I_{n+1} \cong \gamma_n I_{n+1}/I_{n+1} \subseteq I_n/I_{n+1}$. Hence $\gamma_n/\gamma_n \cap I_{n+1}$ is free abelian and $\gamma_n/\gamma_{n+1} \cong \gamma_n \cap I_{n+1}/\gamma_{n+1} \times \gamma_n I_{n+1}/I_{n+1}$. It suffices therefore to prove $\mathrm{rk}(\gamma_n I_{n+1}/I_{n+1}) = \mathrm{rk}(I_n/I_{n+1})$. But this is immediate since $I_n/\gamma_n I_{n+1}$ is periodic, hence finite, and by the preceding remark. The proof is now complete.

Theorem 7. *Let $A = L(X, \mathfrak{V})$ be a free Lie algebra of some Lie algebra variety $\mathfrak{V}$. Then there exists a Lie type variety of groups $\mathfrak{V}^\circ$ such that for any separable filter $\mathcal{F}$ consisting of ideals with nilpotent quotient algebras the subgroup $G = \mathrm{gp}(X)$ of the group $\hat{L}^\circ$, $\hat{L}$ being the completion of L by the filter $\mathcal{F}$, is $\mathfrak{V}^\circ$-free with X as a system of $\mathfrak{V}^\circ$-free generators.*

Proof. Denote by $\mathcal{F}_0$ a filter of A whose members are the members of the lower central series of A. Then the topology determined by $\mathcal{F}_0$ is the strongest one among all the topologies induced by $\mathcal{F}$ described in the theorem. By a remark from section 1 then the completion $\tilde{A}$ by $\mathcal{F}_0$ is isomorphically embedded in $\hat{A}$, the isomorphism continues the identical map of A. By the Campbell-Hausdorff formula and since $X \subseteq \tilde{A}$, one has $G = \mathrm{gp}(X) \subseteq \tilde{A}^\circ$ and thus we may restrict ourselves to the case of $\mathcal{F}_0$.

To prove that G is a free group of some Lie type variety it suffices to show that (i) $\bigcap_{n=1}^{\infty} \gamma_n(G) = \{1\}$ and (ii) any relation in X is an identical relation in A. But, as easily seen, $\gamma_n(G) \subseteq A^n$ (cf. lemma 7.1) so that (i) follows. Further $\tilde{A}/\tilde{A}^n \cong A/A^n$ is a relatively free Lie algebra, hence $(\tilde{A}/\tilde{A}^n)^\circ = \tilde{A}^\circ/(\tilde{A}^n)^\circ$ is a relatively free D-group by the isomorphism of categories (Corollary 8.1). A system of free generators of this group is the image of X under the natural epimorphism. Therefore if $v(x_1, x_2, \ldots, x_n) = 1$ is some relation among $x_1, x_2, \ldots, x_n$, it holds identically in each $\tilde{A}^\circ/(\tilde{A}^n)^\circ$. But $\bigcap_{n=1}^{\infty} (\tilde{A}^n)^\circ = \{1\}$ so that $\tilde{A}^\circ \in \mathrm{var}(\tilde{A}^\circ/(\tilde{A}^n)^\circ, n = 1, 2, \ldots)$ and $v(x_1, x_2, \ldots, x_n) \equiv 1$ is an identity in A. Hence $G = \mathrm{gp}(X)$ is $\mathfrak{V}^\circ$-free for some $\mathfrak{V}^\circ$ and the proof is complete.

Theorem 8. *Given a Lie type variety $\mathfrak{U}$ of groups, there exists a variety $\mathfrak{V}$ of Lie algebras over $\mathbb{Q}$ such that for any nonempty set X, the $\mathfrak{U}$-free group $F(X, \mathfrak{U})$ is isomorphic to the subgroup of $(\hat{L}(X, \mathfrak{V}))^\circ$, the completion of a $\mathfrak{V}$-free algebra $L(X, \mathfrak{V})$, generated by X.*

Proof. Clearly only the case of countable X may be considered. Denote $F(X, \mathfrak{U})$ by G and apply the construction of the final part of the preceding section. Our aim is to show that the subalgebra L of the algebra $\tilde{T}$ (constructed there) generated by X is relatively free, that is $\mathfrak{V}$-free for some variety $\mathfrak{V}$ of Lie algebras over $\mathbb{Q}$. It suffices to prove that any mapping of X into $\tilde{T}$ extends to a homomorphism of L into T or, in other words, that if for some v, $v(x_1, x_2, \dots, x_n) = 0$, then $v(z_1, z_2, \dots, z_m) \equiv 0$ holds in T identically.

Recall that each $x \in X$ (considered as a subset of $\tilde{T}$) has the form $x = \varprojlim x^{(n)}$, $x^{(n)} \in T^{(n)}$, $n = 1, 2, \dots$. Therefore $0 = v(x_1, x_2, \dots, x_m) = \varprojlim v(x_1^{(n)}, x_2^{(n)}, \dots, x_m^{(n)})$. This implies $v(x_1^{(n)}, x_2^{(n)}, \dots, x_m^{(n)}) = 0$ in each $T^{(n)}$. By the proposition 1(i) each $T^{(n)}$ is a relatively free D-group freely generated by the set $X^{(n)}$. Therefore by the isomorphism of categories (Corollary 8.1) $T^{(n)}$ is a relatively free Lie algebra, so that $v(z_1, z_2, \dots, z_m) \equiv 0$ is an identity in $T^{(n)}$. By the construction of the projective limit, $\tilde{T} \in \operatorname{var}(\{T^{(n)}, n = 1, 2, \dots\})$ hence $v(z_1, z_2, \dots, z_n) \equiv 0$ is an identical relation in $\tilde{T}$ and so we have proved that L is $\mathfrak{B}$-free for some $\mathfrak{B}$.

It remains to note that by the Campbell-Hausdorff formula F is a subgroup of the closure $\hat{L}$ of L in T, since $\hat{L}^\circ$ is a subgroup of $\tilde{T}^\circ$ containing X and that the topology of L as a subalgebra of T is determined by the ideals $\tilde{T}_n \cap L$ with nilpotent quotient algebras.

A $\mathfrak{B}$-free Lie algebra L of the just proved theorem may be obtained also in the following way. By proposition 1, in any group $G, f = \{I_n(G)\}_{n\geq 1}$ is a central filtration. Form a graded Lie ring $L^f(G)$ using this filtration (see section 4). Since its homogeneous components are torsion free abelian groups, $L^f(G)$ is a subring of a ring $\mathfrak{L}(G) = \mathbf{Q} \otimes_{\mathbf{Z}} L^f(G)$, which is also a graded ring, whose homogeneous components are the completions of I_n/I_{n+1}, $n = 1, 2, \dots$.

Theorem 9 (Andreev). *Let $F = F(X)$ be a $\mathfrak{U}$-free group of some Lie type variety $\mathfrak{U}$ freely generated by X. Then $M = \mathfrak{L}(F)$ is a $\mathfrak{B}$-free Lie algebra generated by X, $\mathfrak{B}$ being some Lie algebra variety and $(\tilde{M})^\circ \supseteq \operatorname{gp}(X) \cong F$. In other words, denoting $\mathfrak{B}$ by $\mathfrak{L}(\mathfrak{U})$, one has $(\mathfrak{L}(\mathfrak{U}))^\circ = \mathfrak{U}$.*

Proof. We look at F as embedded in $\tilde{T}$ of the previous theorem and let $L = \operatorname{alg}(X)$ in $\tilde{T}$. We show first that $L^n = L \cap \tilde{T}_n$ (see section 1), $n = 1, 2, \dots$ The inclusion $L^n \subseteq L \cap \tilde{T}_n$ being evident, consider a 'Lie' element $w(x_1, x_2, \dots, x_s) \in \tilde{T}_n$. Represent w in the form $w = w_1 + w_2 + \dots + w_{n-1} + w_n + \dots + w_m$, w_i being its homogeneous component of degree i. By our construction, lemma 6,1, corollary 6.3 and since all $T^{(m)}$ are relatively free algebras freely generated by $X^{(m)}$ over a field of characteristic 0 all $w_i \equiv 0$, $1 \leqq i \leqq n - 1$, are identical relations holding in $T^{(n)}$. Since $\tilde{T}$ is 'residually a $T^{(n)}$', $n = 1, 2, \dots, w_1 + w_2 + \dots + w_{n-1} \equiv 0$ in T and therefore in L. Hence $w \in L^n$ and therefore $L^n \supseteq L \cap T_n$.

Now since both L and M are graded algebras and by the definition of bracketing in these algebras, a homomorphism of L into M may be given by the rule

$$w(x_1, x_2, \dots, x_s) \mapsto \tilde{w}(x_1, x_2, \dots, x_s)\, I_{n+1}, \tag{1}$$

where $w(x_1, x_2, \dots, x_s)$ is a linear combination with integer coefficients of some basic monomials of degree n in X, $\tilde{w}(x_1, x_2, \dots, x_s)$ is the corresponding commutator in the group F. By the just proved equality $L^n = L \cap \tilde{T}_n$ this mapping is defined correctly and is injective. To prove its surjectivity, it suffices to notice that it preserves multiplication and that both L and M are generated by X.

Thus the proof is complete.

6.9. Product varieties of groups

A variety $\mathfrak{W}$ of groups is called a *product* of varieties $\mathfrak{U}$ and $\mathfrak{V}$ (denoted by $\mathfrak{U}\mathfrak{V}$) if $\mathfrak{W}$ is a class of groups G possessing a normal subgroup $H \in \mathfrak{U}$ such that $G/H \in \mathfrak{V}$. Assume that $\mathfrak{U}$ is determined by a set U of words and $\mathfrak{V}$ by V, let also F_∞ be a free group of countable rank. Then $\mathfrak{W} = \mathfrak{U}\mathfrak{V}$ is determined by the set $U(V(F_\infty))$ of words. For, $G = F_\infty/U(V(F_\infty)) \in \mathfrak{U}\mathfrak{V}$ (in contradistinction to the Lie algebra case a verbal subgroup of a *normal* subgroup of a group is itself *normal*) and therefore $W(F) \subseteq U(V(F))$. The converse is obvious. Similarly, given any nonempty set X, $F(X, \mathfrak{U}\mathfrak{V}) \cong F(X)/U(V(F(X)))$.

Our purpose in this section is to represent a free group of some rank of a product variety through free groups of the same rank of the factors.

We start with a construction of extreme importance in the theory of group varieties. Let A and B be groups. Form the cartesian product $P = C \prod_{b \in B} A_b$, $A_b \cong A$ for all $b \in B$, and turn P into a B-group by putting $f^{b_1}(b) = f(b_1 b)$, $f \in P$, $b, b_1 \in B$. A split extension S of B by P as a B-group is called a *cartesian wreath product* of A and B and is denoted by $S = A\mathrm{Wr}B$. Thus any element from S has the unique form bf, $b \in B$, $f \in P$, the multiplication being defined by $bf \cdot b_1 f_1 = bb_1 f^{b_1} f_1$.

Proposition 1 (Kalužnin). *Let G be an extension of B by A, that is $G \triangleright A$ and $G/A = B$. Then there exists a monomorphism $\mu\colon G \to S = A\mathrm{Wr}B$ such that $\mu(A) \subseteq P$ and if $\bar{\mu}\colon G/A \to S/P$ is a homomorphism canonically induced by μ, then $\mu = 1_B$.*

Proof. Let $s\colon B \to G$ be an injection such that b^s is a representative of the coset $b \in B$ and $1^s = 1$. Define $\mu\colon G \to S$ by the rule $\mu(g) = \bar{g} f_g$, where $\bar{g} = gA$ and $f_g(b) = ((\bar{g}b)^s)^{-1}\, gb^s$. The reader is suggested to check up that $f_g(b) \in A$ for all g and b. We show next that μ is a homomorphism. Indeed, $\mu(g_1 g_2) = \overline{g_1 g_2}\, f_{g_1 g_2}$, where $f_{g_1 g_2}(b) = ((\overline{g_1 g_2} b)^s)^{-1}\, g_1 g_2 b^s$. On the other hand $\mu(g_1)\, \mu(g_2) = \bar{g}_1 f_{g_1} \bar{g}_2 f_{g_2} = \bar{g}_1 \bar{g}_2 f_{g_1}^{\bar{g}_2} f_{g_2}$. So it suffices to compare the P-components of both sides: $(f_{g_1}^{\bar{g}_2} f_{g_2})(b) = f_{g_1}^{\bar{g}_2}(b)\, f_{g_2}(b) = f_{g_1}(\bar{g}_2 b)\, f_{g_2}(b) = ((\bar{g}_1 \bar{g}_2 b)^s)^{-1} \times \bar{g}_1 (\bar{g}_2 b)^s\, ((\bar{g}_2 b)^s)^{-1}\, g_2 b^s = ((\overline{g_1 g_2} b)^s)^{-1}\, \overline{g_1 g_2} b^s = f_{g_1 g_2}(b)$.

Moreover $\mu(g) = 1$ implies $\bar{g} = 1$, that is $g \in A$. But then $f_g(1) = (\bar{g}1)^s\, g1^s = g$, hence $g = 1$. Thus the proof of the proposition is complete.

Another important definition (similar to that of section 3,6) is one of a *wreath $\mathfrak{V}$-product of groups*. Let $\mathfrak{V}$ be a variety of groups, A be a group from $\mathfrak{V}$, B be any group. A group $W = A\mathrm{wr}_{\mathfrak{V}}B$ is called a *wreath $\mathfrak{V}$-product* of A by B if the following conditions hold:

1. $W = \mathrm{gp}(A, B)$.
2. Let P be the normal subgroup of W generated by A. Then $P \in \mathfrak{V}$.
3. If $\varphi\colon A \to G$, $\psi\colon B \to G$ are two homomorphisms of A, B into a group G such that

 (i) $G = \mathrm{gp}(\varphi(A), \psi(B))$;

 (ii) the normal subgroup T generated by $\varphi(A)$ in G belongs to $\mathfrak{V}$,

then there exists a homomorphism $\chi\colon W \to G$ such that $\chi|_A = \varphi$ and $\chi|_B = \psi$.

Just as in the Lie algebra case the existence of a wreath product is proved by using the free product $\Phi = A * B$ of the groups A and B (the definition rewords that of section 2, 6). Let then Q be a normal subgroup of Φ generated by A, and $V(Q)$ be its verbal subgroup corresponding to $\mathfrak{V}$. Then $V(Q)$ is a normal subgroup of Φ and $\Phi/V(Q)$ is a *split extension* of B by $Q/V(Q)$ (compare with 3,6). Now one immediately verifies the defining properties of $A \operatorname{wr}_{\mathfrak{B}} B$ for $\Phi/V(Q)$.

One of the main results of this section is the

Theorem 10 (Šmel'kin). *Let* $G = F(X, \mathfrak{U}\mathfrak{B})$, $A = F(Y, \mathfrak{U})$, $B = F(Z, \mathfrak{B})$, *where all the sets* X, Y, Z *have the same cardinality, the bijections given by* $x \mapsto y \mapsto z$. *Form a wreath* $\mathfrak{U}$*-product* $W = A \operatorname{wr}_{\mathfrak{U}} B$. *Then the mapping* $x \mapsto yz$ *extends to a monomorphism of* G *into* W.

Proof. By the construction, $W \in \mathfrak{U}\mathfrak{B}$ and thus the mapping $x \mapsto yz$ extends to a homomorphism α of G into W. Consider also $S = V(G) \operatorname{Wr} B$. By the previous proposition there exists a monomorphism $\mu: G \to S$ such that $\mu(V(G)) \subseteq P$. Let $\mu(x) = b_x f_x$, then the mappings $y \mapsto f_x$ and $z \mapsto b_x$ extend to homomorphisms $\varphi: A \to V(G)$ and $\psi: B \to B$. Since P as a cartesian product of groups $\cong V(G)$ is a $\mathfrak{U}$-group, by the definition of a wreath $\mathfrak{U}$-product there exists a homomorphism $\chi: W \to S$ extending φ and ψ. The diagram

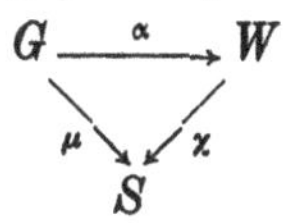

is commutative which proves that α is a monomorphism and completes the proof of the theorem.

One more "Lie type" result also due to A. L. Šmel'kin is the important

Theorem 11. *Let* $L = L(X, \mathfrak{B})$ *be a* $\mathfrak{B}$*-free Lie algebra over* $\mathbb{Q}$, B *be a nilpotent Lie algebra,* $W = L \operatorname{wr}_{\mathfrak{B}} B$ *be their wreath* $\mathfrak{B}$*-product,* $\overline{W}$ *be its completion. Then the subgroup* T *generated by* X *and* B° *in* $\overline{W}^\circ$ *is isomorphic to* $A \operatorname{wr}_{\mathfrak{U}} B^\circ$, A *being a* $\mathfrak{U}$*-free group generated by* X, $\mathfrak{U} = \mathfrak{B}^\circ$.

To clarify the formulation recall, that by the proposition 3,6.11, $\bigcap_{n=1}^{\infty} W^n = \{0\}$ and thus one may speak on the completion of W relatively to the filter formed by these ideals. A closure $\overline{K}$ of $K = \operatorname{id}_W(L)$ in the corresponding topology is its completion by the filter formed by $K_n = K \cap W^n$. By theorem 7, the subgroup A generated by X in $\overline{K}$ is a $\mathfrak{U} = \mathfrak{B}^\circ$-free group. To prove the theorem, notice that by the universal property of $R = A \operatorname{wr}_{\mathfrak{U}} B^\circ$ this is a split extension of B° by a $\mathfrak{U}$-free group S generated by x^b, $x \in X$, $b \in B^\circ$, the action of B° in S being given by $b_1^{-1}(x^b)\, b = x^{bb_1}$, $b \in B$. Thus to prove our theorem it suffices (by theorem 7) to show that the elements x^b, $x \in X$, $b \in B$, form a system of $\mathfrak{B}$-free generators of the subalgebra they generate.

As an exercise using the very definition of the Campbell-Hausdorff series, the reader is proposed to verify, that, given $x, y \in L(\{x, y\}) \subseteq \hat{L}(\{x, y\})$, one has

$$x^{y^{-1}} = yxy^{-1} = \sum_{n=0}^{\infty} \frac{(\operatorname{ad}(y))^n}{n!}(x) = e^{\operatorname{ad}(y)}(x)\,,$$

that is, the inverse of an inner automorphism of $\hat{L}$ as a Lie algebra coincides with that of the group $\hat{L}^\circ$.

By proposition 3,8.1 K/K^2 is a free B- and therefore $U(B)$-module. Clearly then its completion $\overline{K/K^2}$ by the filter of ideals $K^2 + K_n/K^2$, $n = 1, 2, \ldots$, is a free $\overline{U(B)}$-module, where $\overline{U(B)}$ is the completion of $U(B)$ by the filter consisting of the ideals

$U^{(s)}$ (see 3,8.2). To prove the linear independence of the elements x^b, $x \in X$, $b \in B$, it suffices to prove that of e^b for different $b \in B$. This may be done using induction by $\dim(B)$. Let z be a nonzero element of the centre of B and suppose that there exist nonzero $\lambda_1, \lambda_2, \ldots, \lambda_m \in k$ and such that

$$\lambda_1 e^{b_1} + \lambda_2 e^{b_2} + \ldots + \lambda_m e^{b_m} = 0 . \tag{1}$$

Denote by B_1 the quotient algebra B/kz. Then since the canonical mapping $U(B) \to$ $\to U(B_1)$ is continuous, one has

$$\lambda_1 e^{\bar{b}_1} + \lambda_2 e^{\bar{b}_2} + \ldots + \lambda_m e^{\bar{b}_m} = 0 , \tag{2}$$

where $\bar{b}_i = b_i + kz$, $i = 1, 2, \ldots, m$. By the induction hypothesis then one may assume that all $\bar{b}_i$ in (2) coincide and some of λ_i, $i = 1, 2, \ldots, m$, is nonzero. Write therefore $b_i = u + \mu_i z$, then since z is central (2) is reduced to

$$e^u(\lambda_1 e^{\mu_1 z} + \lambda_2 e^{\mu_2 z} + \ldots + \lambda_m e^{\mu_m z}) = 0$$

or

$$\lambda_1 e^{\mu_1 z} + \lambda_2 e^{\mu_2 z} + \ldots + \lambda_m e^{\mu_m z} = 0 . \tag{3}$$

Deriving (3) by z shows that this is possible only whensome of the $\mu_i \in k$ coincide.

It is immediate from here that, given a finite set of elements $x_1^{b_1}, x_2^{b_2}, \ldots, x_n^{b_n} \in \bar{K}$, there exist $s \geqq 1$ and $u_1, u_2, \ldots, u_n$ in K such that $x_i^{b_i} - u_i \in K_s$ and the u_i are linearly independent modulo K^2. By theorem 3,4.2, $u_1, u_2, \ldots, u_n$ are $\mathfrak{B}$-free generators of the subalgebra they generate. Now on one hand by the construction of the completion, $\bar{K} \in \mathfrak{B}$. On the other hand if $v(x_1^{b_1}, x_2^{b_2}, \ldots, x_n^{b_n}) = 0$ is a relation among $x_1^{b_1}, x_2^{b_2}, \ldots, x_n^{b_n}$, then $v(u_1, u_2, \ldots, u_n) = 0$, for, each difference $x_i^{b_i} - u_i$ depends on *other* free generators of K then those of u_i. This shows that $v \equiv 0$ is an identity in $\mathfrak{B}$ and therefore we have proved our theorem.

Corollary. *Let $\mathfrak{U}$ and $\mathfrak{B}$ be two Lie type varieties of groups. Then $\mathfrak{U}\mathfrak{B}$ is also a Lie type variety.*

Proof. By theorem 7 it suffices to show that a wreath $\mathfrak{U}$-product $T = A wr_{\mathfrak{B}} B$ is a Lie type group. If $\mathfrak{B}$ is a nilpotent variety, this follows immediately from the just proved theorem and the already mentioned fact $\gamma_n(\bar{W}^\circ) \subseteq \bar{W}^n$.

The reduction to the nilpotent case may be done as follows. To prove that T is a Lie type group it suffices to find for each $1 \neq a \in T$ a normal subgroup Q_a such that T/Q_a is torsion free nilpotent and $aQ_a \neq 1$. But if $a \notin P$, then $aP \notin I_s(B)$ for some $s > 1$ and therefore $PI_s(B)$ is the desired normal subgroup. Let further $a \in P$. Then $a = w(x_1^{b_1}, x_2^{b_2}, \ldots, x_m^{b_m}) \neq 1$, where $x_i^{b_i}$ are free generators of P. Choose s such that b_i and $b_i b_j^{-1}$ are not elements of $I_s(B)$, $1 \leqq i \neq j \leqq m$, and denote by B_1 the quotient group $B/I_s(B)$. Consider also $T_1 = A\mathrm{wr}_{\mathfrak{U}} B_1$. Then, denoting $bI_s(B)$ by $\bar{b}$, the mapping $x^b \mapsto x^{\bar{b}}$, $b \mapsto \bar{b}$ extends to a homomorphism of T onto T_1 under which $a \mapsto w(x_1^{\bar{b}_1}, x_2^{\bar{b}_2}, \ldots, x_m^{\bar{b}_m}) \neq 1$, since $x_i^{\bar{b}_i}$ are different $\mathfrak{U}$-free generators of P_1. This gives the desired reduction and proves the corollary.

In particular, we can assert that, given a *free solvable group* G of some length 1, one has $\bigcap_{n=1}^{\infty} \gamma_n(G) = \{1\}$. Another group of interest is a *free polynilpotent* group R of polynilpotent length $(n_1, n_2, \ldots, n_s)$, n_i being some positive integers, that is a free group of a product variety $\mathfrak{N}_{n_1}\mathfrak{N}_{n_2} \ldots \mathfrak{N}_{n_s}$. By our corollary also $\bigcap_{n=1}^{\infty} \gamma_n(R) = \{1\}$.

6.10. Magnus groups

A group G is called a *Magnus group* if

(i) $\bigcap_{n=1}^{\infty} \gamma_n(G) = \{1\}$,

(ii) $\gamma_n(G)/\gamma_{n+1}(G)$ is a torsion free group, $n = 1, 2, \ldots$.

The same definition is used for Lie rings, that is Lie algebras over $\mathbf{Z}$. A variety of groups (respectively Lie algebras over $\mathbf{Z}$) is called a *Magnus variety* if its free groups (respectively algebras) are Magnus.

Proposition 1. *Let $\mathfrak{U}$, $\mathfrak{V}$ be Magnus varieties of Lie rings determined by multihomogeneous systems of identical relations. Then their product is a Magnus variety.*

Proof. Proceeding as in the corollary to theorem 3,4.3, one sees that the verbal subring of an absolutely free Lie ring $L(X)$ determining the variety $\mathfrak{U}\mathfrak{V}$ is $U(V(L))$. Clearly it suffices to deal with the case of finite X (for any "torsion" is contained in a subring generated by a finite subset of X). Since $\mathfrak{V}$ is determined by its multihomogeneous identities, $V(L)$ is a homogeneous subring of $L(X)$ such that $L/V(L)$ is torsion free. Now theorem 2,3.4 shows that $V(L)$ is a free subring of L which yields $V(L)/U(V(L))$ to be a Magnus ring. Therefore $L/U(V(L))$ as an abelian group is the extension of the torsion free abelian group $L/V(L)$ by the torsion-free abelian group $V(L)/U(V(L))$ and thus is torsion free. Since, further, $\mathfrak{U}\mathfrak{V}$ is determined by a multihomogeneous system of identities, $T = L/U(V(L))$ is graded, that is $\bigcap_{n=1}^{\infty} T^n = \{0\}$ and $T \cong \bigoplus_{n=1}^{\infty} T^n/T^{n+1}$ (cf. proposition 3,4.2). Thus each T^n/T^{n+1} is torsion free and the proof is complete.

Theorem 12 (Šmel'kin). *Let $\mathfrak{U}$, $\mathfrak{V}$ be two Magnus varieties of groups. Then their product $\mathfrak{U}\mathfrak{V}$ is a Magnus variety.*

We start with a lemma based on the corollary from the previous section which asserts that the product of two Lie type vaiieties of groups is a Lie type one. Notation $\mathfrak{L}(\mathfrak{U})$ see in the theorem 8.9.

Lemma. *Given two Lie type varieties of groups $\mathfrak{U}$, $\mathfrak{V}$, one has $\mathfrak{L}(\mathfrak{U})\,\mathfrak{L}(\mathfrak{V}) = \mathfrak{L}(\mathfrak{U}\mathfrak{V})$.*

Proof. Any Lie algebra variety over a field of zero characteristic is generated by its nilpotent algebras. So far, let $P \in \mathfrak{L}(\mathfrak{U}\mathfrak{V})$. Then $P^\circ \in \mathfrak{U}\mathfrak{V}$ (theorem 8.9). Therefore $P^\circ \rhd Q^\circ \in \mathfrak{U}$ and $P^\circ/Q^\circ = (P/Q)^\circ \in \mathfrak{V}$. By the results of section 7, $Q \in \mathfrak{L}(\mathfrak{U})$ and $P/Q \in \mathfrak{L}(\mathfrak{V})$ which yields $P \in \mathfrak{L}(\mathfrak{U})\,\mathfrak{L}(\mathfrak{V})$. The converse inclusion is similar, and the lemma is proved.

Returning to the proof of the theorem, denote by $L(\mathfrak{V})$ and $L(\mathfrak{U})$ (Magnus) varieties of Lie rings determined by $\mathfrak{V}$ and $\mathfrak{U}$[1]. By the very definition of varieties $L(\mathfrak{U})$, $L(\mathfrak{V})$ their free algebras are graded and hence these varieties are determined by multihomogeneous identities so that the lemma above applies and $L(\mathfrak{U})\,L(\mathfrak{V})$ is a Magnus variety.

[1]) These are varieties whose free rings are subrings of $\mathfrak{L}(\mathfrak{V})$-free algebras freely generated as rings by the same set of free generators. In the particular case of Magnus varieties if G is a $\mathfrak{V}$-free group, $L(G)$ is a $L(\mathfrak{V})$-free ring.

Let $F = F(X)$ be a free Lie algebra of this variety of some finite rank and $G = G(X)$ be a free group of a group variety $\mathfrak{U}\mathfrak{V}$ having the same rank. To complete the proof of our theorem it suffices to show that $L(G) \cong F$ under the canonical homomorphism (if any) induced by the identity map of X. We prove first that $L(G) \in L(\mathfrak{U})\, L(\mathfrak{V})$.

Indeed, note that for each $n \geqq 1$, $\gamma_n(G)/\gamma_{n+1}(G)$ is an extension of $\gamma_n(G)/\gamma_{n+1}(G) \times \times \left(V(G) \cap \gamma_n(G)\right)$ by $\gamma_{n+1}(G)\left(V(G) \cap \gamma_n(G)\right)/\gamma_{n+1}(G)$. The first factor is isomorphic to $\gamma_n\left(G/V(G)\right)/\gamma_{n+1}\left(G/V(G)\right)$, for,

$$\gamma_n(G)/\gamma_{n+1}(G)\left(V(G) \cap \gamma_n(G)\right) = \gamma_n(G)/\gamma_n(G) \cap \left(\gamma_{n+1}(G)\, V(G)\right) \cong$$
$$\cong \gamma_n(G)\, V(G)/\gamma_{n+1}(G)\, V(G) \cong \gamma_n\left(G/V(G)\right)/\gamma_{n+1}\left(G/V(G)\right).$$

The second is isomorphic to $V(G) \cap \gamma_n(G)/V(G) \cap \gamma_{n+1}(G)$. Thus $L(G)$ is an extension of $L\left(G/V(G)\right) \in L(\mathfrak{V})$ by the ideal T constructed from $V(G)$ using the series

$$V(G) \supset V(G) \cap \gamma_2(G) \supset V(G) \cap \gamma_3(G) \supset \dots. \tag{1}$$

But $L\left(V(G)\right) \in L(\mathfrak{U})$ and, for each n, $V(G) \cap \gamma_n(G) \supset \gamma_n\left(V(G)\right)$. This means that any multihomogeneous identity holding in $L\left(V(G)\right)$ holds also in T constructed by (1). But $L(\mathfrak{U})$ is determined by its multihomogeneous identities. Hence $T \in L(\mathfrak{U})$ and $L(G) \in L(\mathfrak{U})\, L(\mathfrak{V})$. Therefore the identity map of X extends to an epimorphism $\varphi\colon F \to \to L(G)$.

For, clearly $F_{\mathbf{Q}} = F \otimes_{\mathbf{Z}} \mathbf{Q}$ is a free algebra of a variety $\mathfrak{L}(\mathfrak{U})\, \mathfrak{L}(\mathfrak{V})$ (F as an abelian group simply the direct sum of two free abelian groups which are free algebras of $L(\mathfrak{U})$ and $L(\mathfrak{V})$. By the lemma, however, $\mathfrak{L}(\mathfrak{U})\,\mathfrak{L}(\mathfrak{V}) = \mathfrak{L}(\mathfrak{U}\mathfrak{V})$ and therefore $F_{\mathbf{Q}}$ is a free algebra of the variety $\mathfrak{L}(\mathfrak{U}\mathfrak{V})$. Hence the n-th homogeneous component of $F_{\mathbf{Q}}$ equals to the completion of $I_n(G)/I_{n+1}(G)$, $n = 1, 2, \dots$, and thus the rank of F^n/F^{n+1} equals (by proposition 8.1) to that of $\gamma_n(G)/\gamma_{n+1}(G)$. Since on the other hand F^n/F^{n+1} is mapped under φ onto $\gamma_n(G)/\gamma_{n+1}(G)$ and $\bigcap\limits_{n=1}^{\infty} F^n = \{0\}$, φ is in fact an isomorphism which completes now the proof of the whole theorem.

Corollary. *Free polynilpotent groups are Magnus ones. In particular free solvable groups are Magnus.*

Sources. References. Discussion

General books on Lie algebras

[1] Amayo, R., Stewart, I., Infinite dimensional Lie algebras. Leyden: Noordhoof 1974.
[2] Bourbaki, N., Groupes et algèbres de Lie. Paris: Hermann 1971, 1972.
[3] Cartier, P., et al., Théorie des algèbres de Lie. Topologie des groupes de Lie. Séminaire ‚Sophus Lie', 1, Paris 1955.
[4] Chevalley, C., Théorie des groupes de Lie, Tome 3. Théorèmes généraux sur les algèbres de Lie. Paris: Hermann 1955.
[5] Jacobson, N., Lie algebras. New York: Interscience 1962.
[6] Kaplansky, I., Lie algebras and locally compact groups, Chicago Lectures in Math. Chicago and London 1971.
[7] Seligman, G., Modular Lie algebras. Berlin/Heidelberg/New York: Springer 1967.
[8] Serre, J. P., Lie algebras and Lie groups. New York: Benjamin 1965.
[9] Stewart, I., Lie algebras, Lecture Notes in Math., v. 127. Berlin/Heidelberg/New York: Springer 1970.

Chapter 1

Section 1. For a Lie algebra structure triviality result see

[10] Nouazé, Y., Un exemple de module sans structure d'algèbre de Lie. C.R. Acad. Sci. Paris, 274 (1972), A158.

Section 2. For connections of Lie algebras with Lie groups see [2] or

[11] Cohn, P. M., Lie groups, Cambridge Tracts etc., v. 46. Cambridge 1957.

Section 3. Classification of simple Lie algebras the in zero characteristic case is carried out in many sources, see e.g. [5] or [8]. One of the shortest expositions is in [9]. The prime characteristic case is the concern of [6] and [7]. A survey paper of interest is

[12] Kostrikin, A. I., Some aspects of Lie algebra theory, 'Selected questions of algebra and logic'. Novosibirsk: Nauka 1973.

Section 9. Engel's theorem originally proved for finite dimensional algebras over fields by now has many generalizations. See e.g.

[13] Amayo, R., Engel Lie rings with chain conditions. Pacific J. Math. 54 (1974), 1–12; and references therein.

Section 12. Engel Lie algebras (that is, satisfying condition $(ad(x))^{n(x)} = 0$ for any x, $n(x)$ being positive) were dealt with by many authors, the most remarkable results being due to Kostrikin and Razmyslov. See also book [1].

[14] Kostrikin, A. I., Lie rings satisfying the Engel condition. Izv. Akad. Nauk SSSR, Ser. Mat. 21 (1957), 515–540.
[15] Razmyslov, Ju. P., On Engel Lie algebras. Alg. i Log. Sem. 10 (1974), 33–44.

A generalization of Higgins' theorem may be found in [17] below.

[16] Higgins, P. J., Lie rings satisfying the Engel condition. Proc. Camb. Phil. Soc. **50** (1954), 8–15.

[17] Gruenberg, K. W., The Engel elements of a soluble group. Illinois J. Math. **3** (1959), 151–168.

Section 13. The Frattini subalgebra was studied originally by D. W. Barnes who proved the theorems 10.5 and 13.8. He also constructed an example of a finite dimensional Lie algebra whose Frattini subalgebra is not an ideal. Further development see in [21] and [20].

[18] Barnes, D. W., Lie algebras. Lecture Notes (University of Tübingen 1968/69).

[19] Barnes, D. W., Newell, M. L., Some theorems on saturated homomorphs of soluble Lie algebras. Math. Z. **115** (1970), 179–187.

[20] Towers, D. A., A Frattini theory for algebras. Proc. London Math. Soc. (3) **27** (1973), 440–462.

[21] Towers, D., Stewart, I., The Frattini subalgebras of certain infinite dimensional soluble Lie algebras. J. London Math. Soc. **11** (1975), 207–215.

Section 15. Representations of Lie algebras and Lie modules are studied in a large number of books and papers. See e.g. [5], [8] or

[22] Humphreys, J., Introduction to Lie algebras and representation theory, Graduate Texts in Math. Berlin: Springer 1972.

[23] Borho, W., Gabriel, P., Rentschler, R., Primideale in Einhüllenden auflösbarer Lie-Algebren, Lecture Notes in Math., v. 357. Berlin/Heidelberg/New York: 1973.

[24] Dixmier, J., Algèbres enveloppantes. Paris: Gauthier-Villars 1974.

General results on modules may be seen in various books, e.g.

[25] Curtis, C., Reiner, I., Representation theory of finite groups and associative algebras. New York/London: Interscience 1962.

Chapter 2

Section 2. The main theorem 1 is due to A. I. Širšov [28]. The exposition given is due to an unpublished report of D. I. Èĭdel'kind. Other related papers and books are listed below:

[26] Hall, M., A basis for free Lie algebras and higher commutators in free groups. Proc. Amer. Math. Soc. **1** (1950), 575–581.

[27] Širšov, A. I., On free Lie rings. Mat. Sb. **45** (1958), 113–122.

[28] Širšov, A. I., On the bases of a free Lie algebra. Alg. i Log. Sem. **1** (1962), 14–19.

[29] Cohn, P. M., Universal algebra. New York/Evanston/London: Harper & Row 1965, chapter 8.6.

[30] Viennot, J., Factorisations réguliers des monoïdes libres et des algèbres de Lie libres. C.R. Acad. Sci. Paris **277** (1973), A493–496.

Hall's type basis is being constructed also in [2], chap. 2.

Section 3. The main theorem is due to Širšov-Witt. The present variant may be seen either in [33] or in [2], chapter 2, § 2, exercises 12, 13. In the paper [31] there is an example showing the failure of the main theorem when the base ring is not a field.

[31] Širšov, A. I., Subalgebras of free Lie algebras. Mat. Sb. **33** (1953), 441–452.

[32] Witt, E., Die Unterringe der freien Lieschen Ringe. Math. Z. **64** (1956), 195–216.

[33] Èĭdel'kind, D. I., On Magnus groups. Mat. Sb. **92** (1973), 209–223.

Section 4. The principal PBW-theorem is proved as in the already quoted book [29]. The proof valid for Lie algebras free as k-modules over any commutative base ring k may be found in [2] or [8]. In the former book (chapter 2, § 2, exercise 9) there is an example of a Lie algebra not embeddable isomorphically in its universal envelope. This example is contained also in Širšov's paper [34] below.

[34] Širšov, A. I., On representation of Lie rings in associative rings. Uspehi Mat. Nauk **8** (1953), 173–175.

Section 5. Theorem 6 is due to E. Witt [32]. Its exposition and theorem 7 are due to G. P. Kukin [35]. In this latter paper see a partial solution of a still unsolved problem: given two finitely generated subalgebras H, K of a free Lie algebra L over some field k; is it true that $H \cap K$ is ditto? Answer: yes, if char $(k) = p > 0$[1]).

[35] Kukin, G. P., On subalgebras of free Lie p-algebras. Alg. i Log. Sem. **11** (1972), 535–550.

Section 6. A basis of a free product of Lie algebras over a field was constructed by A. I. Širšov [36]. For the present state of this subject see Kukin's papers [37], [38].

[36] Širšov, A. I., On a Lie algebra theory conjecture. Sibirsk. Mat. Ž. **3** (1962), 297–301.
[37] Kukin, G. P., Subalgebras of a free sum of Lie algebras with amalgamated subalgebra. Alg. i Log. Sem. **11** (1972), 59–86.
[38] Kukin, G. P., On free products of restricted Lie algebras. Mat. Sb. **95** (1974), 53–83.

Chapter 3

Many results of this chapter are theorems of the group varieties theory treated in

[39] Neumann, Hanna, Varieties of groups. Berlin/Heidelberg/New York: Springer 1967.

Section 3. All results of this section are contained in fact in

[40] Malcev, A. I., On algebras with identical defining relations. Mat. Sb. **26** (1950), 19–23.

Counterexamples showing the necessity of restrictions therein may be found in

[41] Bahturin, Ju. A., On identities in Lie algebras. 1. Vestnik MGU, ser. mat. meh. 1973, no. 1, 12–18.

Sections 4, 5. Proposition 2 is folklore. Theorem 2 is a compound of a result from [41] and a theorem from [42]. Theorem 3 is due to the author [43], [44], and V. A. Parfenov [45]. A theorem on Lie algebra varieties semigroup was proved for zero characteristic case by V. A. Parfenov [46]. The present proof is an improved version of [47]. See also [44] where the groupoid of Lie algebra varieties over the two element field is shown to be nonassociative.

[42] Andreev, K. K., Šabel'nikova, D. G., Characteristic subalgebras of relatively free algebras. Sibirsk. Mat. Ž. **10** (1973), 1336–1337.
[43] Bahturin, Ju. A., Two remarks on Lie algebra varieties. Mat. Zamet. **4** (1968), 387–398.
[44] Bahturin, Ju. A., On identities in Lie algebras, 2. Vestnik MGU, ser. mat. meh. 1973, no. 2, 26–33.
[45] Parfenov, V. A., On a property of ideals in a free Lie algebra. Sibirsk. Mat. Ž. **10** (1969), 940–944.
[46] Parfenov, V. A., On varieties of Lie algebras. Alg. i Log. Sem. **6** (1967), 61–73.
[47] Genov, G. K., On verbal ideals of free Lie algebras. Godisnik Mat. Fak. Sofia. Univ. **66** (1974), 177–189.

Section 6. An original example of an infinitely based variety of Lie algebras was obtained by M.R. Vaughan-Lee [48] for the case of fields of characteristic two. The present theorem 5 follows V. S. Drensky's paper [49]. This theorem was obtained also by Ju. G. Kleiman (unpublished). V. S. Drensky generalises to arbitrary prime characteristic M. R. Vaughan-Lee's result on the existence over any infinite field of characteristic 2 of a finite dimensional Lie algebra having no finite basis of identities. Recently this latter has constructed an example of an infinitely based Lie algebra variety in $\mathfrak{A}\mathfrak{N}_2$ over any characteristic 2 field [50].

[1]) Recently G. P. Kukin informed me that the answer "yes" is true for an arbitrary field.

[48] Vaughan-Lee, M. R., Varieties of Lie algebras. Quart. J. Math. Oxford 21 (1970), 297—308.
[49] Drensky, V. S., On identities in Lie algebras. Alg. i Log. Sem. 13 (1974), 265—290.
[50] Vaughan-Lee, M. R., Abelian-by-nilpotent varieties of Lie algebras. J. London Math. Soc. 11 (1975), 263—266.

Sections 7, 8. Wreath products of Lie algebras were introduced and applied to group theory by A. L. Šmel'kin [51]. A theorem on finite basis of identities in metabelian varieties is contained in M. R. Vaughan-Lee's thesis (Oxford, 1968); it apes D. E. Cohen's machinery from [52]. The partially well ordered set technique due to G. Higman [53] enables to prove that, for any $c \geqq 1$, any subvariety of $\mathfrak{A}\mathfrak{N}_c \cap \mathfrak{N}_c\mathfrak{A}$ and any subvariety of $\mathfrak{N}_2\mathfrak{A}$ is finitely based (in the latter case the main fiel dis assumed as having characteristic $\neq 2$) [54][1]. Recall that the finite basis problem is not yet solved when char $(k) = 0$. To this end mention a profound result due to Ju. P. Razmyslov [55], showing any subvariety of var $(sl_2(k))$ finitely generated.

[51] Šmel'kin, A. L., Lie algebra wreath products and their application in group theory. Trudy Moskov. mat. obšč. 29 (1973), 247—260.
[52] Cohen, D. E., On the laws of a metabelian variety. J. Algebra 5 (1967), 267—273.
[53] Higman, G., Ordering by divisibility in abstract algebras. Proc. London Math. Soc. 2 (1952), 326—336.
[54] Bryant, M. R., Vaughan-Lee, M. R., Soluble varieties of Lie algebras. Quart. J. Math. Oxford 23 (1972), 107—112.
[55] Razmyslov, Ju. P., Finite basis properties in certain varieties of algebras. Alg. i Log. Sem. 13 (1974), 685—693.

Chapter 4

Section 1. Primary facts on PI-algebras see in [56], [57], or in [58]. SPI-algebras were studied by V. N. Latyšev [59]. He pointed out an example of a Lie PI-algebra which is not SPI.

[56] Jacobson, N., Structure of rings. Amer. Math. Soc. Providence 1964.
[57] Jacobson, N., PI-algebras. An introduction. Lecture Notes in Math. v. 441. Berlin/Heidelberg/New York: Springer 1975.
[58] Herstein, I., Noncommutative rings. Carus monograph, Wiley & Sons 1968.
[59] Latyšev, V. N., Algebras with identical defining relations. Sibirsk. Mat. Ž. 4 (1963), 821—829.

Section 2. A result on identities in the universal envelopes in characteristico is due to

[60] Latyšev, V. N., Two remarks on PI-algebras. Sibirsk. Mat. Ž. 4 (1963), 1120—1121.

Section 3. Theorem 2 was proved by C. W. Curtis. For deeper results on representations see references cited in 1, 15.

[61] Curtis, C. W., Noncommutative extensions of Hilbert rings. Proc. Amer. Math. Soc. 4 (1953), 945—955.

Sections 4—7. Theorem 3 is due to P. M. Neumann [62]. The remaining part of the section 4 as well as sections 5—7 are formed by the author's results from [63]. S. Amitsur's theorem referred to in section 5 may be found in [64]. Its more general version is also available in the already quoted [58].

[62] Neumann, P. M., An improved bound for BFC-groups. J. Austral. Math. Soc. 11 (1970), 19—27.

[1]) These results were generalized by G. V. Šeina (Sibirsk. Mat. Ž. 17 (1976), 194—199) who showed that any subvariety of $\mathfrak{N}_c\mathfrak{A} \cap \mathfrak{N}_2\mathfrak{N}_c$ is finitely based provided that the characteristic of the base field is not 2.

[63] Bahturin, Ju. A., Identities in the universal envelopes of Lie algebras. Mat. Zamet. 18 (1974), 10–21.
[64] Amitsur, S. A., On rings with identities. J. London Math. Soc. 30 (1955), 464–470.

Section 8. Theorems and examples relevant to residual finiteness of Lie algebras were obtained first in author's [65]. See also [66].

[65] Bahturin, Ju. A., On residual properties of Lie algebras. Mat. Zamet. 12 (1972), 719–723.
[66] Amayo, R., Finiteness conditions in soluble Lie algebras. J. London Math. Soc. 8 (1974), 283–289.

Section 9 consists of results from

[67] Bahturin, Ju. A., Ol'šanskiĭ, A. Ju., Identical relations in finite Lie algebras. Mat. Sb. 96 (1975), 543–559.

Chapter 5

This chapter consists entirely of the results obtained in [67]. For comparison with analogous results in other classes of universal algebras see the survey paper

[68] Macdonald, S. O., Various varieties. J. Austral. Math. Soc. 16 (1973), 363–367.

In particular, the group theoretical origin may bee seen in [69] or in the book [39]. For associative ring theoretical parallels see [70] and [71].

[69] Oates, S., Powell, M. B., Identical relations in finite groups. J. Algebra 1 (1964), 11–39.
[70] L'vov, I. V., On associative rings with identities 1. Alg. i Log. Sem. 12 (1973), 262–267.
[71] Kruse, R., Identities satisfied by a finite ring. J. Algebra 26 (1973), 298–318.

Further development is achieved in

[72] Bahturin, Ju. A., Ol'šanskiĭ, A. Ju., Solvable just-non-Cross varieties of Lie algebras. Mat. Sb. 100 (1976), 384–399,

where the entitled varieties are completely described. A variety generated by a finite Lie algebra is frequently called *Cross* variety. Suppose that $\mathfrak{V}$ is a variety all of whose subvarieties but $\mathfrak{V}$ itself are Cross. Then $\mathfrak{V}$ is called "just-non-Cross". The description of unsolvable just-non-Cross varieties in contradistinction to the solvable case is incomplete. Their existence however is proved in Razmyslov's paper [15].

Chapter 6

Sections 1–6 contain well known results. Other approaches to the Campbell-Hausdorff formula and the Magnus embedding of free groups into a power series ring may be found in [2], [5], [8]. See also [73] below which may serve as a source of the lacking group theoretical material as well as [74] or [39]. Malcev completion has several approaches listed below ([75], [76], [77], [78], [79]). Using lemma 6.3 for this purpose is due to A. L. Šmel'kin.

[73] Magnus, W., Karras, A., Solitar, D., Combinatorial group theory. New York/Sydney/London: Interscience 1966.
[74] Hall, M., The theory of groups. New York: Macmillan 1959.
[75] Malcev, A. I., Nilpotent torsion free groups. Izv. Akad. Nauk SSSR, ser. mat. 13 (1949), 201–212; On normed Lie algebras over the rational number field. Dokl. Akad. Nauk SSSR 62 (1948), 745–748.
[76] Lazard, M., Sur les groupes nilpotents et les anneaux de Lie. Ann. Sci. École Norm. Supér. 71 (1954), 101–190.

[77] Šmel'kin, A. L., Nilpotent products and torsion free nilpotent groups. Sibirsk. Mat. Ž. 3 (1962), 625–640.
[78] Baumslag, G., Lecture notes on nilpotent groups. Amer. Math. Soc. Providence 1972.
[79] Stewart, I., An algebraic treatment of Malcev's theorems concerning nilpotent Lie groups and their Lie algebras. Compos. Math. 22 (1970), 189–312.

Sections 7, 8. follow to K. K. Andreev's papers (cf. also [82]).

[80] Andreev, K. K., Nilpotent groups and Lie algebras. Alg. i Log. Sem. 7 (1968), 4–14.
[81] Andreev, K. K., Nilpotent groups and Lie algebras 2. Alg. i Log. Sem. 8 (1969), 625–635.
[82] Èidel'kind, D. I., On exact representations of relatively free groups. Alg. i Log. Sem. 10 (1971), 449–473.

In the latter paper the mapping $\mathfrak{B} \to \mathfrak{B}^0$ is shown noninjective.

Sections 9, 10 are entirely due to A. L. Šmel'kin [51]. Group verbal wreath products are also due to the same author ([83], [84]). In the paper [85] A. L. Šmel'kin shows that the free product of Magnus groups is itself Magnus. Magnus groups were studied thoroughly by D. I. Èidel'kind [33]. A particular case of free polynilpotent groups was exhausted also by purely group theoretical methods by K. W. Gruenberg [86]. Connected with this subject are also papers [87], [88], [89], [90].

[83] Šmel'kin, A. L., Wreath products and varieties of groups. Izv. Akad. Nauk SSSR, ser. mat. 29 (1965), 149–170.
[84] Šmel'kin, A. L., A remark to my paper "Wreath products and varieties of groups". Izv. Akad. Nauk SSSR, ser. mat. 31 (1967), 443–444.
[85] Šmel'kin, A. L., On free products of Magnus groups. Alg. i Log. Sem. 8 (1969), 129–137.
[86] Gruenberg, K. W., Residual properties of infinite soluble groups. Proc. London Math. Soc. 7 (1957), 29–62.
[87] Šmel'kin, A. L., Free polynilpotent groups. Izv. Akad. Nauk SSSR, ser. mat. 28 (1964), 91–122.
[88] Kikodze, È. B., On free groups of certain varieties. Alg. i Log. Sem. 5 (1966), 15–23.
[89] Gorčakov, Ju. M., On central series of free groups of some varieties. Alg. i Log. Sem. 6 (1967), 15–23.
[90] Gorčakov, Ju. M., Multinilpotent groups. Alg. i Log. Sem. 6 (1967), 25–30.

Author Index

Amayo, R. 117, 121
Amitsur, S. 67, 120
Andreev, K. K. 111, 119, 121

Bahturin, Ju. A. 46, 72, 74, 76, 119, 120, 121
Barnes, D. W. 11, 15, 118
Baumslag, G. 121
Birkhoff, G. 33, 40
Borho, W. 118
Bourbaki, N. 117
Bryant, M. R. 120

Cartier, P. 117
Cayley, A. 64
Chevalley, C. 117
Cohen, D. E. 56, 120
Cohn, P. M. 117, 118
Curtis, C. W. 63, 118, 120

Dixmier, J. 118
Drensky, V. S. 50, 119, 120

Èidel'kind, D. I. 118, 122
Engel 9, 10, 13

Gabriel, P. 118
Genov, G. K. 47, 119
Gorčakov, Ju. M. 122
Gruenberg, K. W. 118, 122

Hall, M. 118, 121
Hall, Ph. 2, 100
Hamilton, W. R. 64
Herstein, I. 120
Higgins, P. J. 12, 13, 118
Higman, G. 58, 120
Hilbert, D. 64
Humphreys, J. 118

Iwasawa 60

Jacobson, N. 117, 120

Kalužnin, L. A. 112
Kaplansky, I. 63, 117
Karras, A. 121
Kikodze, È. B. 122
Kleiman, Ju. G. 119
Kostrikin, A. I. 117
Kruse, R. 121
Kukin, G. P. 35, 119

Latyšev, V. N. 63, 120
Lazard, M. 121
L'vov, I. V. 121

Macdonald, S. O. 121
Magnus, W. 121
Malcev, A. I. 106, 119, 121

Neumann, Hanna 119
Neumann, P. M. 65, 120
Newell, M. L. 118
Nouazé, Y. 117

Oates, S. 121
Ol'šanskiĭ, A. Ju. 76, 121

Parfenov, V. A. 48, 119
Poincaré, H. 33
Powell, M. B. 121

Razmyslov, Ju. P. 117, 120, 121
Reiner, I. 118
Rentschler, R. 118

Šabel'nikova, D. G. 119
Schur, I. 64
Šeina, G. V. 120
Seligman, G. 117
Serre, J. P. 117
Širšov, A. I. 26, 27, 30, 37, 118, 119
Šmel'kin, A. L. 56, 113, 115, 120, 121, 122
Solitar, D. 121
Stewart, I. 117, 118, 121

Towers, D. A. 16, 118

Vaughan-Lee, M. R. 119, 120
Viennot, J. 118

Witt, E. 30, 33, 34, 118, 119

Subject Index

A-module, fully reducible 19
A-module, irreducible 19
A-module, simple 19
A-submodule 18
abelian Lie algebra 2
abelian socle 15
algebraic Lie algebra 76
algebraic transformation 71

basic monomial 24, 27, 34, 37
basic set 24
basic word 27
Birkhoff's theorem 40
breadth of an element 65

Campbell-Hausdorff formula 97
canonical map 31
cartesian product of Lie algebras 5
Cauchy sequence 92
centre-by-metabelian Lie algebra 74
characteristic subalgebra 7, 44
central filtration 100
centralizer 14
centre 4
chief factor 3
chief series of *A*-submodules 19
closure 57
commutator in a group 99
commutator in a Lie algebra 1
complement of an ideal 16
complete space 92
completion of an algebra 92
composition series 9
converging sequence 92
critical Lie algebra 81
Cross variety 121

D-group 104
degree 22, 29
dense subset 92
derivation 2
derivation, inner 4
derived algebra 7
derived series in a group 100
derived series in a Lie algebra 7
direct product of Lie algebras 5

elementary transformation 34
Engel Lie algebra 117
Engel's theorem 10
equivalent symbols 106
equivalent words 98
extension 11

factor closed set 83
factor of a Lie algebra 3
filtered topological algebra 92
finite basis property 57
finitely based variety of Lie algebras 42
formula $\mathcal{F}^{v}_{d,s}$ 163
Frattini ideal 11
Frattini subalgebra 9
free algebra 22
free associative algebra 23
free generators 22
free group 98
free Lie algebra 23
free Lie-p-algebra 33
free Lie ring 30
free nilpotent (*D*-)group 107
free polynilpotent group 114
free product of associative algebras 37
free product of Lie algebras 36
free restricted Lie algebra 33
free solvable algebra 56
fully invariant subalgebra 41
fully reducible *A*-module 19
fully reducible series 20, 83

generated as an ideal 3
generating set 3
generators, free 22
graded Lie algebra 29
graded Lie ring 100
group, *D*- 104
group, free 98

group, free nilpotent (D-) 107
group, free polynilpotent 114
group, Lie type 109
group, Magnus 115
group, monolithic 103
group, nilpotent 102
group of a Lie algebra 96
group, relatively free 109
group, residually torsion free nilpotent 108
group, $\mathfrak{B}$-free 109
groupoid 22

Hall's basis 27
Higgins' theorem 13
homogeneous element 22, 29
homomorphism 3
homomorphism, p- 17

ideal 3
ideal, Frattini 11
ideal, p- 17
ideal, Φ- 57
ideal, proper 3
identical relation 39
indecomposable variety of Lie algebras 45
infinite sum 94
initial segment 58
inner automorphism 56
inner derivation 4
invariant subalgebra 44
irreducible A-module 19
irreducible word 98
isolated subgroup, isolator 104
isomorphism 3

Jacobi identity 1
junior part of an element 34

length of a module 9
Lie algebra 1
Lie algebra, abelian 2
Lie algebra, algebraic 76
Lie algebra, centre-by-metabelian 74
Lie algebra, critical 81
Lie algebra, Engel 117
Lie algebra, free 23
Lie algebra, free restricted 33
Lie algebra, graded 29
Lie algebra, locally finite 41
Lie algebra, metabelian 12
Lie algebra, monolithic 73
Lie algebra, nilpotent 8
Lie algebra, relatively free 41
Lie algebra, residually finite 73
Lie algebra, restricted 17
Lie algebra, simple 3
Lie algebra, solvable 12
Lie algebra, $\mathfrak{B}$-free 41
Lie p-algebra 17
Lie p-algebra, free 33
Lie ring 1
Lie ring, free 30
Lie ring, graded 100
Lie ring of a group 101
Lie type group 109
Lie type variety 109
linearization of an identity 42
locally finite Lie algebra 41
lower central series in a group 100
lower central series in a Lie algebra 7
$\mathfrak{L}(\mathfrak{B})$ variety of Lie algebras 111

Magnus algebra 94
Magnus group 115
Magnus variety 115
major part of an element 30, 34
Malcev completion 107
metabelian Lie algebra 12
minimal representation 83
module of finite length 9
module over a Lie algebra 18
monolith of a group 103
monolith of a Lie algebra 73
monolithic group 103
monolithic Lie algebra 73
monomial 22
monomial basic 24, 27, 34, 37
multidegree 22
multihomogeneous element 22
multihomogeneous subspace 23
multilinear element 22

nilpotency class 8
nilpotent group 102
nilpotent Lie algebra 8
nilpotent of index n 60
nilpotent radical, nilradical 15
nongenerator 9
normalizer 8, 191

p-algebra, free Lie 33
p-algebra, Lie 17
p-basic monomial 34
p-homomorphism 17
p-ideal 17
p-independent set 34
p-reduced set 34
p-subalgebra 17
partially well ordered set 58
PBW-theorem 32
Φ-ideal 57
Φ-subalgebra 57
Φ-submodule 57
PI-Algebra 59

pivotal element 58
polynomial identity 59
product of groups, wreath (cartesian) 112
product of groups, wreath $\mathfrak{V}$- 112
product of Lie algebras, cartesian 5
product of Lie algebras, direct 5
product of Lie algebras, free 36
product of Lie algebras, semidirect 5
product of Lie algebras, subdirect 70, 73
product of Lie algebras, wreath $\mathfrak{V}$- 53
product of two varieties 45, 112
projective limit 93
proper ideal 3
proper subalgebra 3

quotient algebra 3
quotient module 18

radical, nilpotent 15
radical, solvable 15
rank 109
relatively free group 202
relatively free Lie algebra 41
representation of a Lie algebra 18
residual property 73
residually finite Lie algebra 73
residually torsion free nilpotent group 108
restricted Lie algebra 17
right quotient field 67

S-irreducible 28
S-reducible 28
semidirect product of Lie algebras 5
simple A-module 19
simple Lie algebra 3
Širšov's basis 27
solvable Lie algebra 12
solvable radical 15
SPI-algebra 59
split extension 5
standard identity 61
stronger topology 92
subalgebra 3
subalgebra, characteristic 7, 44
subalgebra, Frattini 9
subalgebra, fully invariant 41
subalgebra, invariant 44
subalgebra, p- 17
subalgebra, Φ- 57
subalgebra, proper 3
subalgebra, verbal 41
subdirect product of Lie algebras 70, 73

terminal segment 58
topological algebra 92

universal enveloping algebra 31
universal properties 23
upper central series in a group 100
upper central series in a Lie algebra 7

var (Qs-1) A variety of Lie algebras 81
variety of groups 109
variety of groups, Lie type 109
variety of groups, Magnus 115
variety of Lie algebras 40
variety of Lie algebras, Cross 121
variety of Lie algebras, finitely based 42
variety of Lie algebras, indecomposable 45
variety of Lie algebras, $\mathfrak{L}(\mathfrak{V})$ 111
variety of Lie algebras, Magnus 115
variety of Lie algebras, var (Qs-1)A 81
variety of Lie algebras, $\mathfrak{V}^{(n)}$ 86
variety of Lie algebras, $\mathfrak{V}^{\circ}$ 110
verbal subalgebra 41
$\mathfrak{V}^{(n)}$ variety of Lie algebras 86
$\mathfrak{V}^{\circ}$ variety of Lie algebras 110
$\mathfrak{V}$-free group 109
$\mathfrak{V}$-free Lie algebra 41

Witt algebra 2
word 41
wreath (cartesian) products of groups 112
wreath $\mathfrak{V}$-product of groups 112
wreath $\mathfrak{V}$-product of Lie algebras 53

Bei Fragen zur Produktsicherheit wenden Sie sich bitte an:
If you have any questions regarding product safety, please contact:

Walter de Gruyter GmbH
Genthiner Straße 13
10785 Berlin
productsafety@degruyterbrill.com